99 ELECTRONIC PROJECTS

by

Herbert Friedman

HOWARD W. SAMS & CO., INC.
THE BOBBS-MERRILL CO., INC.
INDIANAPOLIS • KANSAS CITY • NEW YORK

FIRST EDITION

THIRD PRINTING—1973

International Standard Book Number: 0-672-20818-0
Library of Congress Catalog Card Number: 77-143036

CONTENTS

Project *Page*

PHOTOGRAPHY PROJECTS

POWER SUPPLIES

RADIO RECEIVERS

REMOTE CONTROLS

TEST EQUIPMENT (GENERAL)

TEST EQUIPMENT (AUDIO)

TEST EQUIPMENT (RF)

TRANSMITTER PROJECTS

MISCELLANEOUS PROJECTS

INTRODUCTION

In the past, the electronics hobbyist needed to remember only a handful of useful, practical vacuum-tube circuits in order to assemble countless projects for various applications in his home, job, automobile, and hobby. However, the transistor and other solid-state devices changed all of this. No longer could circuits that could be applied in almost countless situations be memorized. Not only the circuits themselves but the individual or discrete components within the circuits changed over wide latitudes for different semiconductors. Thus, the need materialized for a good reference book on solid-state circuits.

An important part of the presentation of the projects in this book is the availability of the circuit components. Admittedly, resistors and capacitors are easily obtained, or substitutions can be made within limited ranges. However, far too often, semiconductor types are expensive and difficult to obtain. In fact, many industrial types can be obtained only on quantity orders of 25 or more from specialized parts houses. Therefore, the circuits shown in this book were selected because the solid-state components are of the hobby type. These components are readily available over the counter at most electronic parts stores, or they can be ordered from mail-order houses at reasonable prices.

Construction details are provided where necessary. If there are no specific instructions, the circuit can be built in any manner and in any cabinet, provided reasonable care is taken. When metal cabinets must be used, we tell you so. Otherwise, you can use the conventional black plastic case sold by most parts houses, or even pill bottles and plastic butter dishes. Do not worry at all about the heat sinks. When heat sinks are required, we tell you about it; otherwise, simply ignore them.

To make things as easy as possible, capacitor symbols in the schematics have either a straight line and a curved line, or a straight line set in a bracket. The first symbol indicates a nonpolarized capacitor and you can install it without regard to any markings because there is no polarity to be observed. The capacitors indicated by the line and bracket are polarized electrolytic capacitors and must be wired for the polarity shown. The line in the bracket is marked with a plus sign, and the polarity must be double-checked, because the project probably will not work if the capacitor connections are reversed. In fact, serious damage to the capacitor and other circuit components may occur when the polarity is incorrect and power is applied.

Some capacitor voltage ratings might seem excessive, such as a 500-volt disc specified for a 9-volt circuit. In all instances, we have specified the lowest cost capacitor. A 500-volt disc would cost less than, say, a 10-volt miniature capacitor. Since electrolytic capacitors often represent the biggest expenditure for a project, we suggest you use the lowest cost units you can get whenever possible. When a capacitor value is critical, we specify a silver mica type. The minimum silver mica voltage rating you can easily obtain is 100 volts—use this rating to obtain the lowest cost. To be on the safe side, never use a capacitor with a voltage rating lower than that specified.

Potentiometers can be any taper unless a particular taper is specified. When batteries are specified, do not use a smaller size than recommended. It is much wiser to stick to the original recommendation. Current requirements for a project are taken into account for the battery type suggested in the parts list.

Every circuit will work with the specified transistors, but there is a normal variation in transistor characteristics that can affect performance. For example, a 2N3391 transistor has a possible gain range of 250 to 500, a 2-to-1 difference. If the unit you use has a gain of 500, the base bias becomes critical, and the specified bias resistor might not work in your project. If you have an amplifier that distorts at high levels or an oscillator that will not start, try changing the value of the base bias resistor. This resistor is the one that is connected directly to the base terminal of the transistor and to the collector or to the collector supply. Vary it approximately 20 percent both above and below the recommended value while seeking optimum performance.

It does not matter whether or not you are an experienced electronics hobbyist, your success with the circuits in these projects is assured. Proceed with enthusiasm; your reward will be enjoyment, as well as deep personal satisfaction.

AUDIO PROJECTS

1 SPEAKER MICROPHONE PREAMPLIFIER

A speaker can often serve as a microphone in intercoms, one-way telephones, or as an emergency microphone. All the speaker needs is amplification to raise the speaker's "voice power" output voltage to normal microphone level.

A small speaker-microphone preamplifier can easily be constructed from spare components and just about any general purpose transistor with a beta (current gain) of 30 to about 150. While a pnp transistor is shown, an npn type can be substituted if capacitor C1 and the battery polarities are reversed. No other changes are needed.

Q1 is a common-base amplifier providing a low impedance input to match low impedance speakers of 3.2, 4, 6, 8, and 16

PARTS LIST

Item	Description
B1	Battery, 9-volt, Type 2U6 (or equiv)
C1	Capacitor, electrolytic, 6-μF, 25-Vdc
C2	Capacitor, 0.47-μF, 10-Vdc
Q1	Transistor, pnp, GE-2 (or equiv)
R1	Resistor, 270,000-ohm, ½-watt
R2	Resistor, 27,000-ohm, ½-watt
S1	Switch, spst
SP1	Speaker, pm, 4 to 16 ohms

ohms. The output is medium impedance. The 0.47 μF capacitor (C2) allows the preamplifier to work into loads of 7000 ohms or higher with maximum low-frequency response.

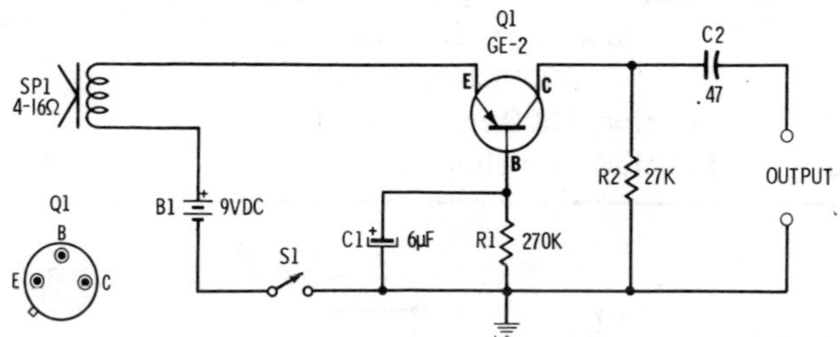

Fig. 1. This circuit will make a high impedance microphone out of one of your speakers.

2 CARBON MICROPHONE CONVERTER

Just as good pitching beats good hitting, a good magnetic microphone beats a good carbon microphone. This one-transistor carbon microphone converter allows a magnetic microphone to be connected in place of a carbon microphone.

Note that no common ground connection is used, even if the circuit is built in a metal cabinet. M1 is a replacement-type magnetic element that is substituted for the original carbon element. Using miniature components, the converter amplifier can be housed in the original microphone case. To avoid damaging transistor Q1, the unit must be connected properly the first time. The positive (+) lead, connects to the collector of Q1, and must be connected to the carbon microphone input lead that supplies a positive voltage.

PARTS LIST

Item	Description
C1	Capacitor, electrolytic, 10-μF, 10-Vdc
M1	Microphone, magnetic replacement element
Q1	Transistor, npn, 2N3394
R1	Resistor, 2200-ohm, ½-watt
R2	Resistor, 6800-ohm, ½-watt
R3	Resistor, 240-ohm, ½-watt

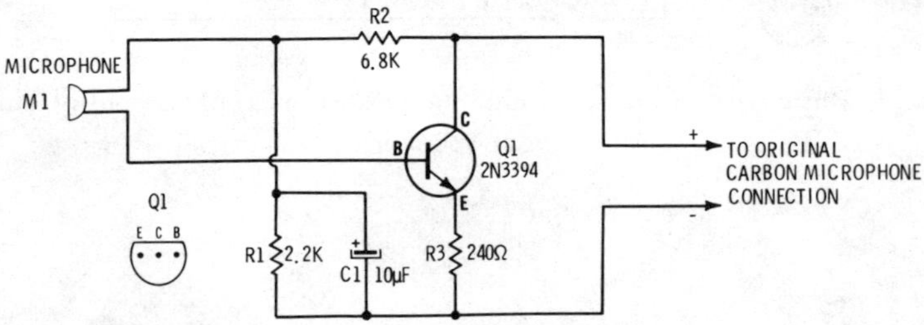

Fig. 2. Exchange your carbon microphone for a magnetic microphone with the use of this circuit.

3 CERAMIC PICKUP PREAMPLIFIER

Because of record equalization and the normal characteristics of a ceramic pickup, the sound reproduced by a "flat" amplifier can be lacking in lows (high-pitched). Some sort of equalization is needed when feeding the pickup to, say, a public-address system, tape recorder, or auxiliary amplifier.

This one-transistor preamplifier provides proper equalization for modern ceramic pickups, and amplification to allow direct connection to a high-level input.

Equalization is accomplished through low impedance loading of the pickup by the input impedance of R1 and Q1, and by the equalizer network (C1, C2, R3). Changes in equalization can be obtained by adjusting the equalizer network values. Do not change the value of R2 for equalization, although it can be adjusted to correct the base bias to accommodate other transistor types or substitutes.

PARTS LIST

Item	Description
C1	Capacitor, 0.01-μF, 10-Vdc
C2	Capacitor, 0.0033-μF, 100-Vdc
C3	Capacitor, 0.05-μF for tube amplifier; 10-μF, 10-volt for transistor amplifier
C4	Capacitor, electrolytic, 30-μF, 6-Vdc
Q1	Transistor, npn, Motorola HEP-641 (or equiv)
R1	Resistor, 10,000-ohm, ½-watt
R2	Resistor, 220,000-ohm, ½-watt
R3	Resistor, 33,000-ohm, ½-watt
R4	Resistor, 4700-ohm, ½-watt

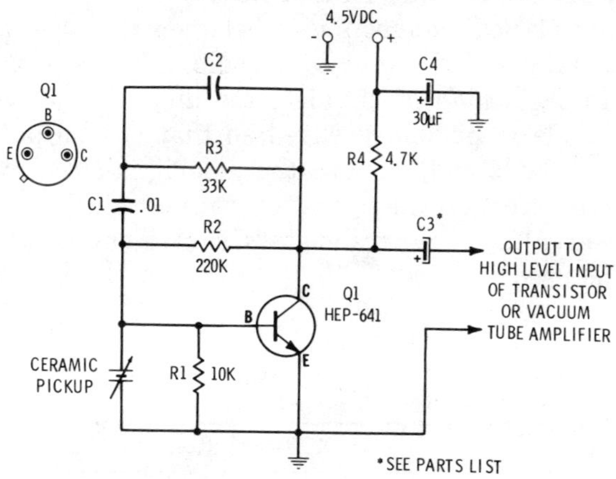

Fig. 3. This circuit will help to get the low tones back in your sound.

4 CRYSTAL RADIO AMPLIFIER

Even with spare parts, this three-stage output transformerless (OTL) amplifier will produce table-radio volume from a simple crystal detector radio.

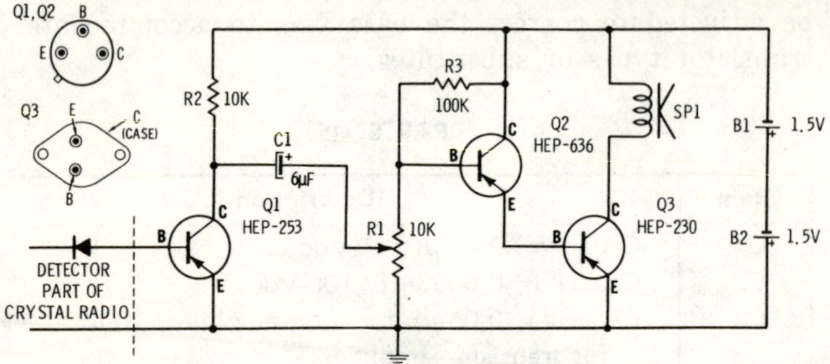

Fig. 4. That crystal detector radio can be heard by everyone with the addition of this amplifier.

Note that an unusual connection for volume control R1 is used because the end-to-end resistance of R1 is part of the base-bias divider for Q2 (R1 and R3).

The only critical connection is between the base of Q1 and the diode detector of the crystal radio. The base of Q1 must connect to the *anode* of the detector diode, as shown. If the detector diode is presently wired so that the output is taken from the cathode end, reverse the polarity of the diode. This will have no effect on the operation of the radio.

Any general purpose transistors equivalent to those specified can be used.

PARTS LIST

Item	Description
B1, B2	Battery, 1.5-volt "D" cell
C1	Capacitor, electrolytic, 6-µF, 6-Vdc
Q1	Transistor, pnp, Motorola HEP-253 (or equiv)
Q2	Transistor, pnp, Motorola HEP-636 (or equiv)
Q3	Transistor, pnp, Motorola HEP-230 (or equiv)
R1	Potentiometer, 10,000-ohm
R2	Resistor, 10,000-ohm, ½-watt
R3	Resistor, 100,000-ohm, ½-watt
SP1	Speaker, 3.2-ohm

5 HI-Z MICROPHONE PREAMPLIFIER

Approximately 10 dB of extra microphone amplification for CB or ham transmitters, tape recorders, and pa amplifiers is provided by the field-effect transistor (FET).

Since the input impedance of the FET is many megohms, the input impedance of the amplifier is determined by the gate resistor (R1), which is 2 megohms—a suitable load for high impedance crystal and ceramic microphones.

The amplifier is "flat" from 20 to 20,000 Hz. Low frequencies can be attenuated for communications use by reducing the value of capacitor C2 to 0.002 μF.

PARTS LIST

Item	Description
C1, C2	Capacitor, 0.05-μF, 25-Vdc
C3	Capacitor, electrolytic, 100-μF, 15-Vdc
Q1	FET transistor, Motorola MPF-103 (or equiv)
R1	Resistor, 2-megohm, ½-watt
R2	Resistor, 3300-ohm, ½-watt
R3	Resistor, 10,000-ohm, ½-watt

Power supply bypass capacitor C3 must be used regardless of whether the power supply is a rectifier or battery. If C3 is eliminated, there might be severe low frequency attenuation, reduced gain, or instability.

The output of the amplifier can be connected to any load of 50,000 ohms or greater.

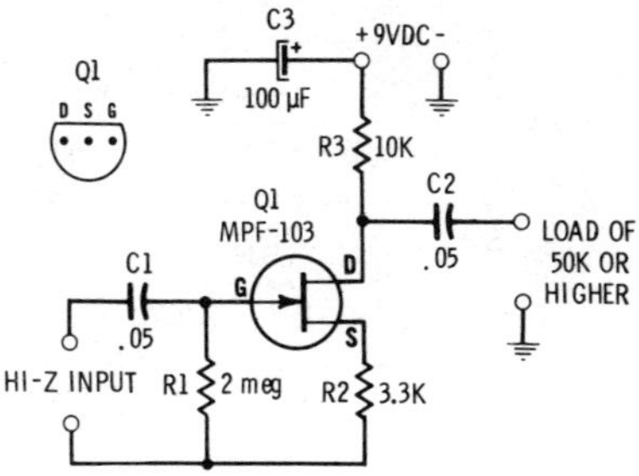

Fig. 5. The transistor used in this circuit is called a junction field-effect transistor (JFET).

6 BUDGET MICROPHONE MIXER

This two-channel microphone mixer will handle any dynamic microphone, either high or low impedance. It can also be used as an emergency mixer for crystal and ceramic microphones although it will produce some low frequency attenuation.

PARTS LIST

Item	Description
B1	Battery, 9-volt, Type 2U6 (or equiv)
C1, C2	Capacitor, 0.1-μF, 6-Vdc
C3	Capacitor, electrolytic, 10-μF, 15-Vdc
Q1	Transistor, pnp general-purpose, 2N107, 2N109, etc.
R1, R2	Potentiometer, audio taper, 2-megohm, ½-watt
R3, R4	Resistor, 100,000-ohm, ½-watt
R5	Resistor, 15,000-ohm, ½-watt
S1	Switch, spst

Transistor Q1 can be any general purpose type such as the 2N107 or 2N217. Although higher quality transistors will provide a better signal-to-noise ratio, the top quality, high-gain transistors should not be used since the relatively high leakage current of "experimenter grade" transistors provides the base bias current. Transistors with low leakage might produce high distortion because of inadequate base bias.

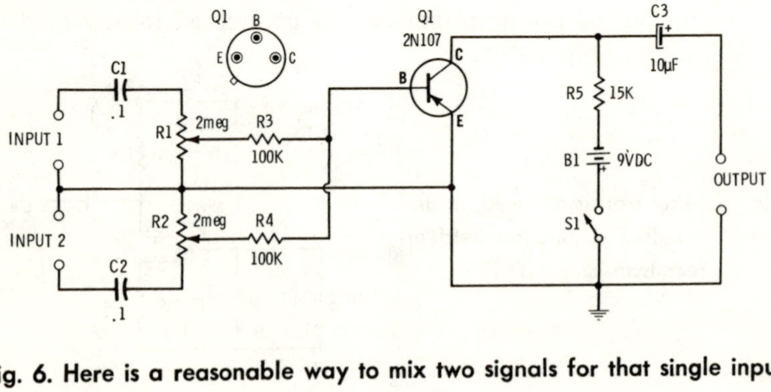

Fig. 6. Here is a reasonable way to mix two signals for that single input.

7 HI-Z AMPLIFIER MICROPHONE MIXER

For serious recording of anything other than speech and sound effects, two microphones are always better than one. The amplified microphone mixer maintains sound quality by compensating for the mixer loss through preamplification. This provides a higher signal-to-noise ratio compared to the simplified mixers that mix first and then amplify.

PARTS LIST

Item	Description
C1, C4	Capacitor, 0.05-μF, 16-Vdc
C2, C5	Capacitor, electrolytic, 25-μF, 25-Vdc
C3, C6	Capacitor, 0.1-μF, 25-Vdc
Q1, Q2	FET transistor, Motorola HEP-801 (or equiv)
R1, R6	Resistor, 2-megohm, ½-watt
R2, R7	Resistor, 6800-ohm, ½-watt
R3, R8	Resistor, 560-ohm, ½-watt
R4, R9	Potentiometer, audio taper, 500,000-ohm
R5, R10	Resistor, 100,000-ohm, ½-watt

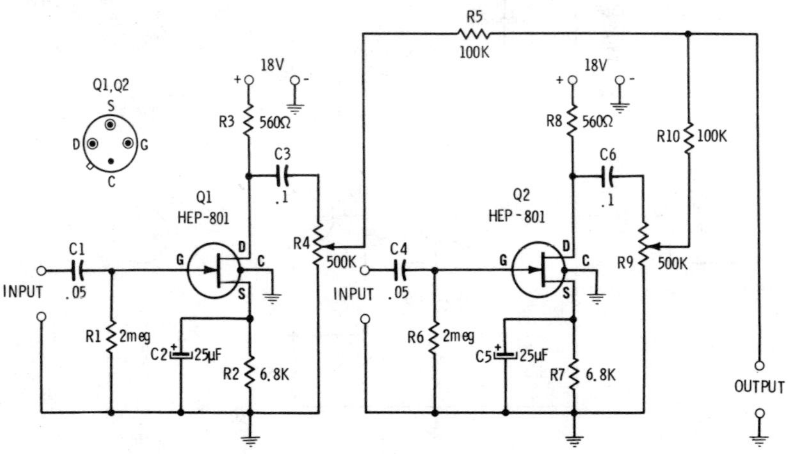

Fig. 7. These high-gain mixers draw only a few milliamperes from the battery.

Because the input impedance into Q1 and Q2 is 2 megohms (the value of R1 and R6), the mixer can be used with crystal and ceramic microphones. There is no loss of low frequency response caused by low impedance loading. The overall frequency response is "flat" from 20 to 20,000 Hz.

Two mixers can be built into the same cabinet for stereo use. Even with two independent mixers (for stereo), current drain is in the order of a few milliamperes, and two series-connected, transistor-radio type batteries can be used.

8 LINE-POWERED PHONOGRAPH AMPLIFIER

Vacuum tube type phonographs are easily updated with this 1-watt output, line-powered transistor amplifier.

No power transformer is needed because Q1 and Q2 are designed to work directly from a high-voltage power source. Full power output is produced by phonograph pickups with a rated 0.5-volt output level. If a "low impedance" ceramic pickup is

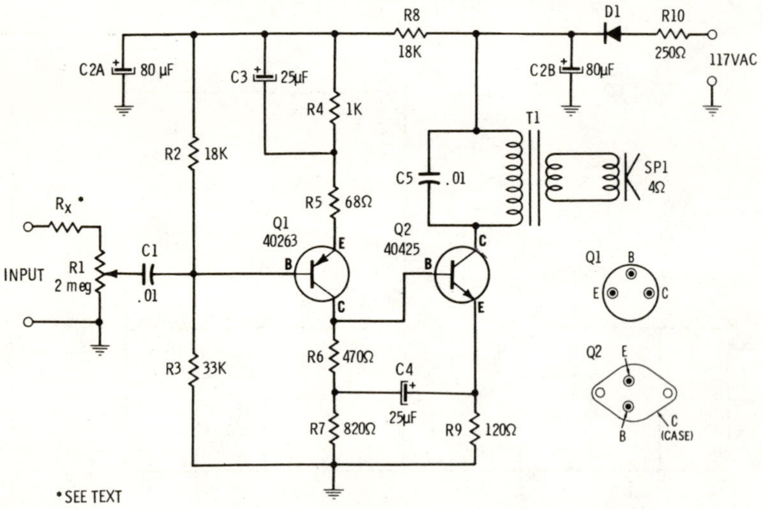

Fig. 8. Notice that one side of the 117-volt line is connected to the circuit ground.

used, eliminate the series input resistor. If a standard high impedance pickup is used, install a resistor having a value of 300,000 to 500,000 ohms for R_x.

Optimum sound balance can be obtained by slight changes in the value of C5. A higher value attenuates the high frequencies, and a lower value increases the high frequencies. Tone control circuits, if desired, should be installed ahead of volume control R1.

PARTS LIST

Item	Description
C1	Capacitor, 0.01-μF, 200-Vdc
C2	Capacitor, electrolytic, 80/80-μF, 6-Vdc
C3, C4	Capacitor, 25-μF, 6-Vdc
C5	Capacitor, 0.01-μF, 400-Vdc
D1	Rectifier, silicon, 750-mA, 400-PIV
Q1	Transistor, pnp, RCA 40263 (or equiv)
Q2	Transistor, npn, RCA 40425 (or equiv)
R1	Potentiometer, audio taper, 2-megohm
R2, R8	Resistor, 18,000-ohm, ½-watt
R3	Resistor, 33,000-ohm, ½-watt
R4	Resistor, 1000-ohm, ½-watt
R5	Resistor, 68-ohm, ½-watt
R6	Resistor, 470-ohm, ½-watt
R7	Resistor, 820-ohm, ½-watt
R9	Resistor, 120-ohm, ½-watt
R10	Resistor, 250-ohm, 1-watt
R_x	See text
SP1	Speaker, 4-ohm
T1	Transformer, audio output, 2500-ohm primary, 4-ohm secondary
Misc.	Heat sink for transistor Q2

CAUTION: Since one side of the power line provides the ground connection, for maximum safety make certain no wire or component connected to "ground" is exposed, and that no ground connection is made to the top deck of the phonograph.

9 LO-Z MICROPHONE PREAMPLIFIER

Just a handful of parts is all it takes to add up to 30-dB gain for low-impedance microphone inputs found on tape recorders, CB rigs, etc. The circuit is suitable for mikes in the 50- to 1000-ohm range.

PARTS LIST

Item	Description
B1	Battery, 6V, four "AA" or "AAA" cells
C1	Capacitor electrolytic 10-μF, 15-Vdc
C2	Capacitor 0.47-μF
Q1	Transistor, npn, 2N3391
R1	Resistor, 10,000-ohm, ½-watt
R2	Resistor, 15-ohm, ½-watt
R_x	See text

Because transistor Q1 is a high-gain type it is sensitive to slight changes in base bias. Hence, bias resistor R_x must be tailored for each transistor. Temporarily connect a 2-megohm potentiometer in place of R_x and adjust the potentiometer until the collector to ground voltage is 3 volts. Measure the potentiometer resistance and substitute a fixed resistor that is within 10 percent of the measured value.

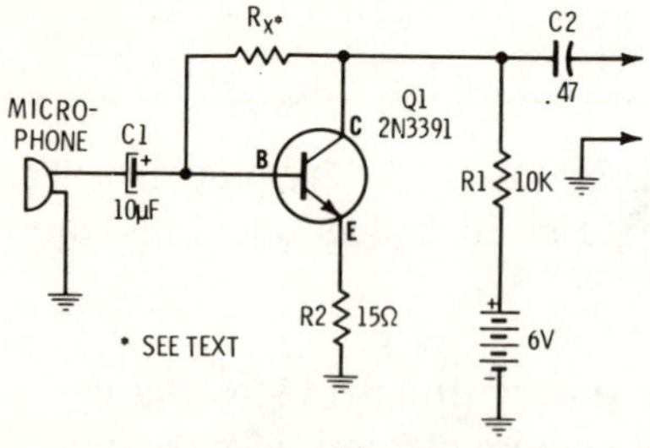

Fig. 9. Improve the gain of that low impedance microphone.

10 POWER MEGAPHONE

This power megaphone is suitable for use at boats, playing fields, meetings, etc. Just about any power transistor having ratings similar to the 2N301 type can be used for Q1 and Q2.

The transistors are connected in parallel to handle the required power and to match the speaker impedance. The microphone is a carbon type such as the ones used in a telephone handset. If a surplus carbon microphone is used, the push-to-talk (PTT) switch can be connected in place of S1 to provide PTT operation.

PARTS LIST

Item	Description
B1, B2	Battery, 6-volt lantern or four "D" cells
M1	Microphone, carbon
Q1, Q2	Transistor, pnp, Motorola HEP-230 (or equiv)
R1	Potentiometer, 5000-ohm
SP1	Speaker or horn, 4-ohm
S1	Switch, spst (see text)

Batteries B1 and B2 are 6-volt lantern type or four "D" cells in holders. Two lantern batteries or eight "D" cells are required. The unit should be built in a metal cabinet, with the cabinet serving as the transistor heat sink. Use mica insulators

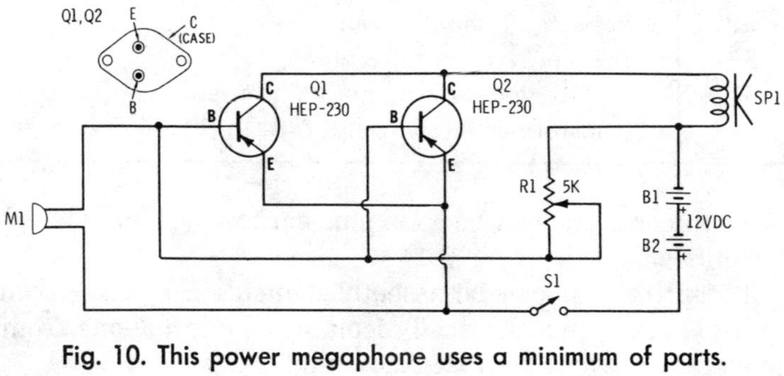

Fig. 10. This power megaphone uses a minimum of parts.

coated on both sides with silicone grease between each transistor and the cabinet.

Potentiometer R1 is adjusted for maximum sound output consistent with lowest distortion.

There is no warm up or "capacitor charge" time; the unit is ready as soon as switch S1 is closed.

11 LOUDHAILER

Although the design is simple and easy-to-build, this one-transistor loudhailer puts out a powerhouse shout. The circuit, except for the microphone, can be assembled in a metal cabinet with a paging horn or trumpet speaker mounted on top of the cabinet.

Transistor Q1 must be provided with a heat sink, which can be the cabinet itself. Take care, however, that the case of Q1—the collector—is insulated from the cabinet. Mounting hardware is available in a transistor mounting kit.

PARTS LIST

Item	Description
B1	Battery, 6-volt lantern or four "D" cells
M1	Carbon microphone
Q1	Transistor, pnp, Motorola HEP-232 (or equiv)
R1	Resistor, 270-ohm, ½-watt
R2	Resistor, 1-ohm, 4-watt
SP1	Speaker, horn-type, 8-ohm
S1	Push-button switch, normally open
T1	Transformer, audio output, 8- to 20-ohm

The microphone can be a surplus carbon type or a telephone transmitter.

The entire unit can be assembled inside a speaker-trumpet if care is taken to acoustically isolate the microphone from the speaker to prevent feedback howling.

Note that transformer T1 must be rated for at least 5 watts. Do not use a miniature transistor transformer.

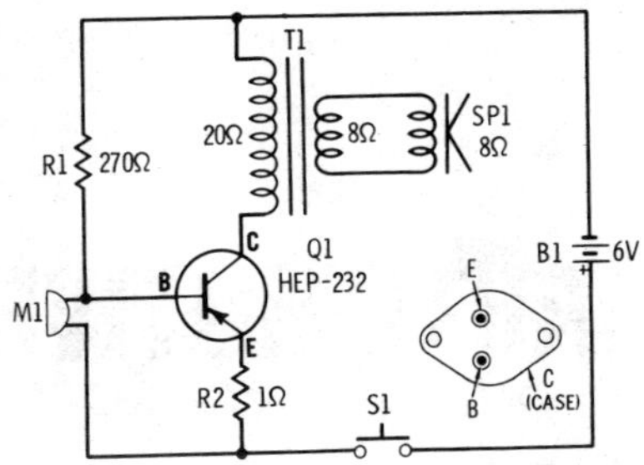

Fig. 11. A loud sound is produced by this one-transistor amplifier.

AUTOMOTIVE PROJECTS

12 AUTO TACHOMETER

This tachometer, which is easily built and calibrated, is accurate enough to be used for engine tune-ups. Wiring is not critical, and the unit can be assembled in a plastic or metal cabinet. Zener diode D1 is any 250-milliwatt type rated as close as possible to 9 volts.

This tachometer can be used only on cars with a negative battery ground. The positive (+12 volts) input connects to any 12-volt positive point in the wiring; the "–" (ground) input connects to the chassis. The distributor input (R1) connects to the lead between the distributor and ignition coil. Do not

PARTS LIST

Item	Description
C1	Capacitor, electrolytic, 1-μF, 100-Vdc
C2	Capacitor, 0.47-μF, 150-Vdc
D1	Zener diode, 9.1-volt, 250-mW
D2, D3	Rectifier, silicon, 100-mA, 50 PIV
M1	Meter, 0-1 mA dc
Q1	Transistor, npn, RCA SK3020 (or equiv)
R1	Resistor, 200-ohm, ½-watt
R2	Resistor, 220-ohm, ½-watt
R3	Resistor, 1500-ohm, ½-watt
R4	Resistor, 330-ohm, ½-watt
R5	Potentiometer, 1000-ohm

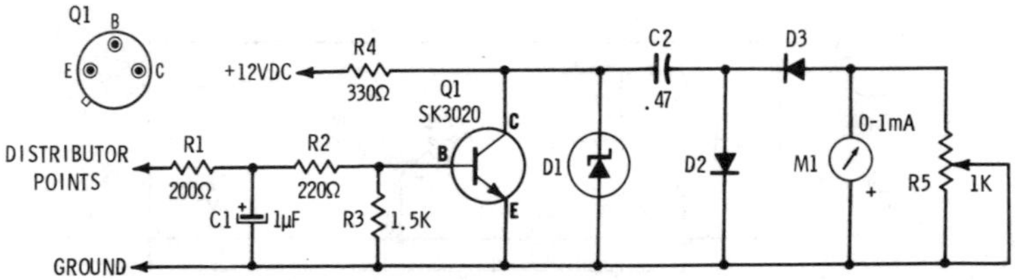

Fig. 12. Check your engine rpm with this reliable tachometer.

use this unit on vehicles equipped with a solid-state ignition system.

The meter scale is linear, with full scale representing approximately 10,000 rpm. Calibrate this tachometer against a commercial one (at your local garage) by adjusting R5 until meter M1 indicates the same reading.

13 TENNA-BLITZ

The ballgame is over, and your car is buried in the parking lot along with about two thousand other cars of the same color. However, you can easily find your car if you have installed a Tenna-Blitz. Sticking above acres of metal is a little lamp going "blink-blink-blink."

Mount the No. 49 pilot lamp at the top of the radio antenna and run two wires down to the control unit inside the car. When switch S1 is closed, turning on the Q1/Q2 multivibrator, the lamp will blink away. The blink rate can be changed by changing the value of capacitor C1.

Almost any general-purpose transistors can be substituted for Q1 and Q2.

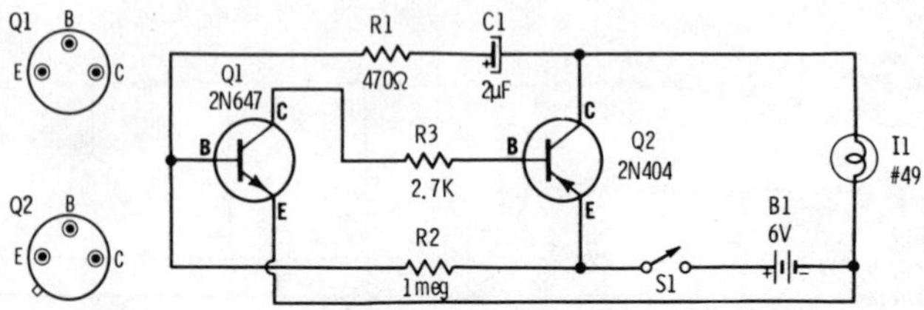

Fig. 13. Here is a device to help you find your car.

PARTS LIST

Item	Description
B1	Battery, 6-volt
C1	Capacitor, electrolytic, 2-μF, 10-Vdc
I1	Pilot lamp, No. 49
Q1	Transistor, npn, 2N647
Q2	Transistor, pnp, 2N404
R1	Resistor, 470-ohm, ½-watt
R2	Resistor, 1-megohm, ½-watt
R3	Resistor, 2700-ohm, ½-watt
S1	Switch, spst

14 AUTO ADAPTER FOR TRANSISTOR RADIOS

When your auto radio stops working, this regulated voltage adapter will keep your portable radio powered until you are ready to work on the radio.

Power is taken from the 12-volt auto battery through the cigarette lighter plug. The series resistor (R1) drops the battery voltage and the zener diode (D1) maintains a 9-volt output for the transistor radio. Zener diode D1 can be any reference or regulator diode with a voltage rating of approximately 9 volts. For example, you can use a 9.1-volt unit (common in zener diode kits) or even one rated at 8.6 volts. Make certain the zener diode is correctly installed so that the end marked with a band (the cathode) connects to resistor R1.

The adapter is rated for a maximum of 12 mA. A good rule of thumb is that a radio powered by a Type 2U6 battery (or equiv) can be safely powered by this adapter.

PARTS LIST

Item	Description
C1	Capacitor, 0.05-μF, 400-Vdc
D1	Diode, zener, 9-volt, 1-watt
PL1	Cigarette lighter plug
R1	Resistor, 150-ohm, ½-watt

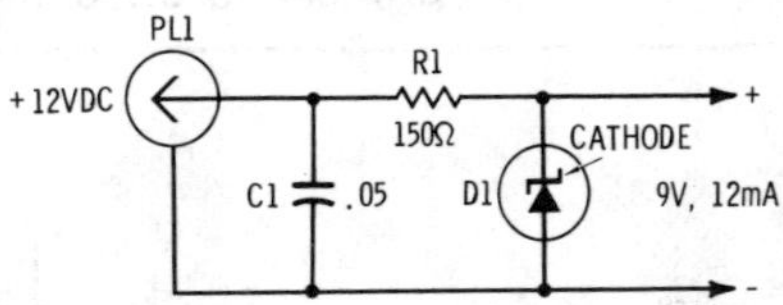

Fig. 14. With this circuit you can use that short-wave or fm portable in your car.

15 HEADLIGHT MINDER

This headlight minder emits a loud and clear signal if you leave the lights on after you turn off the ignition.

When only the ignition is on, the tone circuit is off, since there is no complete power path for Q1. When the lights and the ignition are on, both collector and emitter of Q1 are positive, so the tone generator remains off.

When the lights are on and the ignition is off, the collector of Q1 is placed at battery-negative through R1, and the generator sounds off, letting you know you have left the lights on.

The unit can be assembled in a metal cabinet that is fastened to the dashboard. R1 will be connected through the cabinet to the auto chassis, thereby completing the negative battery connection.

PARTS LIST

Item	Description
C1	Capacitor, electrolytic, 30-μF, 25-Vdc
C2	Capacitor, 0.2-μF, 25-Vdc
D1	Rectifier, silicon, 500 mA, 50-PIV
Q1	Transistor, pnp, RCA SK3005 (or equiv)
R1	Resistor, 15,000-ohm, ½-watt
R2	Resistor, 680-ohm, ½-watt
S1	Switch, dpdt
SP1	Speaker, 3.2-ohm
T1	Transformer, audio output, 500-ohm center-tapped primary to 3.2-ohm secondary

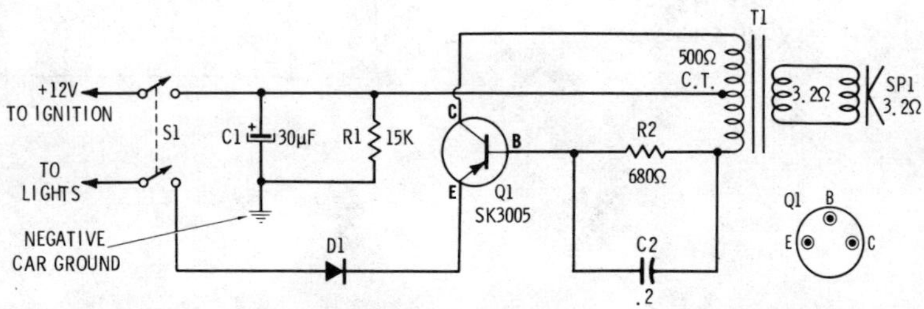

Fig. 15. This circuit will let you know if your headlights are on when the ignition is turned off.

BURGLAR-INTRUDER ALARMS

16 ELECTRONIC SIREN

Put this unit in a portable case with a horn or speaker on top and it can be used as a portable siren whose pitch is varied by pumping the momentary contact switch S2.

Almost any type of construction will work. The overall tone quality can be adjusted by changing the value of C2. If the siren pulsates before S2 is activated, Q1 has excessive leakage, so try a different transistor of the same type.

PARTS LIST

Item	Description
B1	Battery, 6-volt or 12-volt
C1	Capacitor, electrolytic, 30-μF, 15-Vdc
C2	Capacitor, 0.02-μF, 75-Vdc
Q1	Transistor, npn, Motorola HEP-641 (or equiv)
Q2	Transistor, pnp, Motorola HEP-242 (or equiv)
R1, R2	Resistor, 56,000-ohm, ½-watt
R3	Resistor, 27,000-ohm, ½-watt
S1	Switch, spst
S2	Push-button switch, normally open
SP1	Speaker or horn, 8-ohm

When used as a *screamer* for burglar protection, S2 is replaced with a wire jumper, and a normally open door switch is wired in place of switch S1.

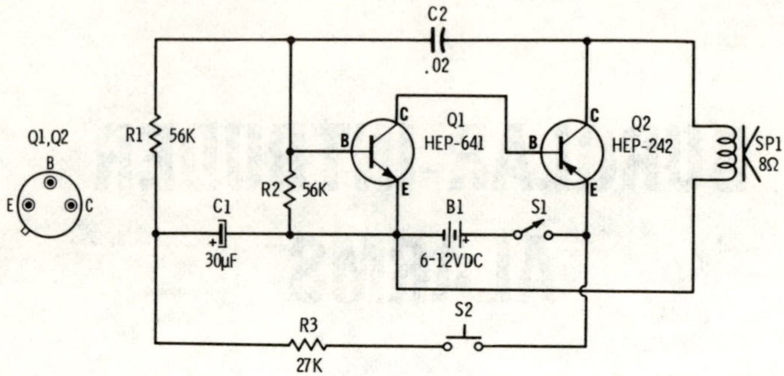

Fig. 16. This circuit can be used as a siren, or as the alarm in a burglar protection system.

17 FLOOD ALARM

If you use this flood alarm, you will not have a worry about water in the basement ruining your electronic gear.

Somewhere near the water pipes, or a low spot in the floor, position two wires spaced approximately one-inch apart. Secure the two wires so they cannot be moved easily. Place about one teaspoon of table salt between the wires. If the floor is con-

PARTS LIST

Item	Description
D1	Diode, 1N60
K1	Relay, 300-ohm, 6-Vdc, P & B Type RS-5D-6 (or equiv)
Q1	Transistor, npn, 2N3393
R1	Potentiometer, 2-megohm
R2	Resistor, 22-ohm, ½-watt

crete, mount the wires and the salt on a sheet of plastic, since the salt can affect the concrete.

When water contacts the salt, the salt becomes a conductor. This completes the base-bias circuit of transistor Q1 and causes relay K1 to be closed by the collector current of transistor Q1. The contacts of K1 close the circuit to an alarm bell.

To adjust the circuit, apply water to a test mound of salt and turn potentiometer R1 until relay K1 closes.

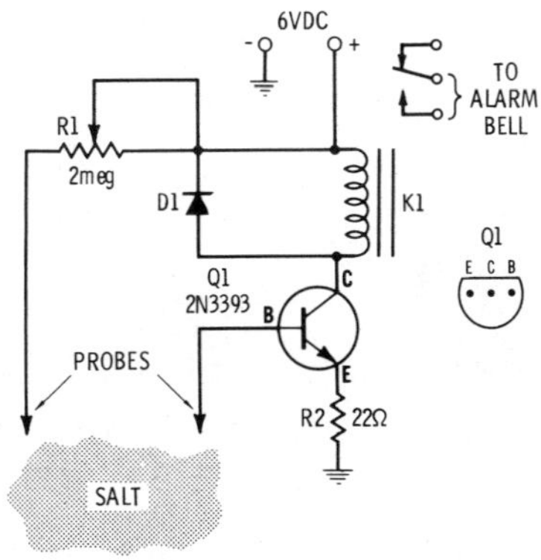

Fig. 17. This circuit makes use of the change in electrical conduction of salt. When salt is dry it is a poor conductor, but when it is wet it is a good conductor.

18 LATCHING BURGLAR ALARM

Open a fancy, and expensive, commercial burglar alarm and all you will find inside is an ordinary latching relay circuit.

The input terminals are connected to parallel-wired, normally open (N.O.) switches. When a security switch closes the series-connected battery circuit, relay K1 pulls in and is "latched" by one set of contacts. The second set of contacts completes the alarm bell circuit. Even if the security switches are disabled, the latching contacts on relay K1 keep the relay closed.

Switch S2 disables or resets the relay circuit and should be in a concealed location to prevent tampering.

PARTS LIST

Item	Description
B1	Battery, 6-volt lantern
M1	Alarm bell, 6-Vdc
K1	Relay, 6-Vdc, dpdt
S1	Switch, spst, normally open
S2	Switch, spst

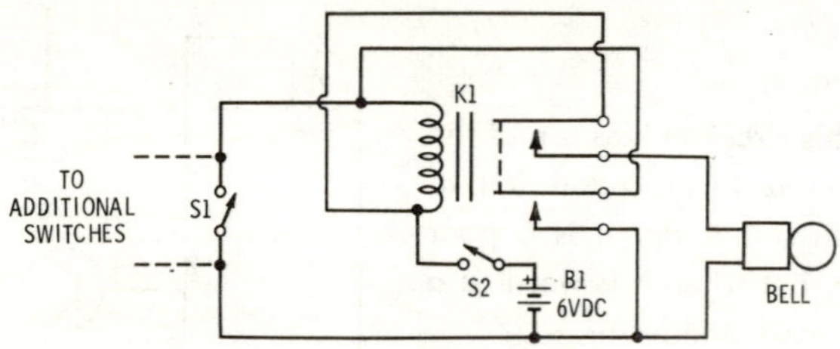

Fig. 18. This alarm will continue to ring until it is reset.

19 POWER FAILURE ALARM

Never fear again that you will be late for work or an appointment because a power failure knocked out the electric alarm clock. The instant the electricity fails, the power failure alarm signals you, even in the early hours of the morning.

To keep current consumption, and therefore operating costs, at rock-bottom, control relay K1 is a "sensitive" 3-mA type, such as is used for radio-control devices.

As long as the power is on, relay K1 is activated and the contacts controlling the buzzer circuit are held open. If power

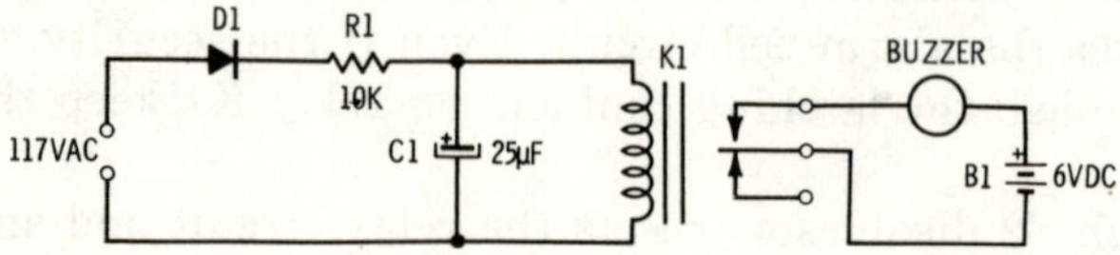

Fig. 19. When power fails, the buzzer will sound until power is restored.

fails, the contacts of relay K1 spring back, closing the battery-to-buzzer circuit.

PARTS LIST

Item	Description
B1	Battery, 6-volt lantern
C1	Capacitor, electrolytic, 25-μF, 150-Vdc
D1	Rectifier, silicon, 500-mA, 200-PIV
K1	Relay, 3000- to 5000-ohm (see text)
R1	Resistor, 10,000-ohm, ½-watt
Misc.	Buzzer or bell, 6-Vdc

20 ELECTRONIC COMBINATION LOCK

Install an electronic combination lock on the ignition system of your car, and a thief would have a better chance playing Russian roulette with a fully loaded six-shooter.

Tracing the circuit in Fig. 20 shows that the siren (or other alarm device) is disabled only if switches S2 and S4 are down and switches S1, S3, and S5 are up. The siren sounds if any other arrangement than the one shown is set when the ignition is turned on. A simple wiring change allows you to set any combination.

All switches are spdt, rather than a few spst, to keep all external switch markings uniform. It would be a dead give-away if two keying notches or lettering positions were reversed.

The switches can be "sporty" auto-accessory types sold individually or in switch banks. If you provide labels such as "Carburetor Heater," "Window Washer," etc., no one will know the car is wired for sound.

PARTS LIST

Item	Description
S1 to S5	Switch, spdt (see text)

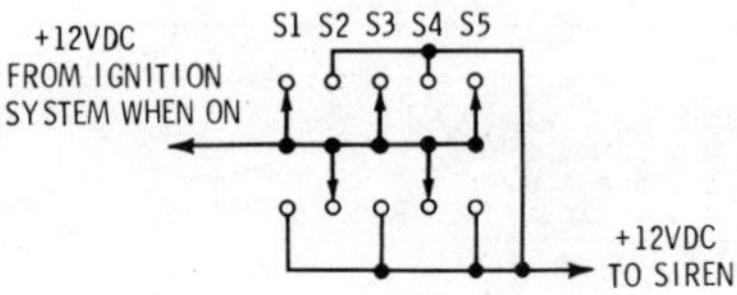

Fig. 20. With the switches in this position, the circuit to the siren is open. All other combinations will set off the siren.

CB, HAM, AND SWL PROJECTS

21 BROADCAST BAND BOOSTER

You can dig out signals you never knew existed from under the "dead spots" of your broadcast band (BC) radio with this one-evening project that delivers an overall circuit gain of some 3 to 5 S-units.

Antenna coil L1 is connected to the circuit using the terminal arrangement given in the instructions packed with the coil. Capacitor C1 can be any 365-pF tuning capacitor such as one salvaged from an old radio, or a low-cost *poly* type will be all right. The radio-frequency choke (RC) must be no larger than 1 mH to avoid instability and oscillation.

The connection from the booster can be made through a plug and jack. However, the connecting cable, can be made of RG-174 coaxial cable, and must be no longer than 12 inches, or there will be severe signal attenuation.

The high gain of the booster can cause oscillation if the output cable gets near the antenna. For best results, assemble the booster in a metal cabinet and ground the cabinet to a cold water pipe. If the BC receiver is the ac/dc type, connect a 0.05-μF, 400-volt capacitor between the output cable shield and the "ground" of the receiver. When using the FET in a circuit, be sure to observe the precautions listed under the Photo of the FET.

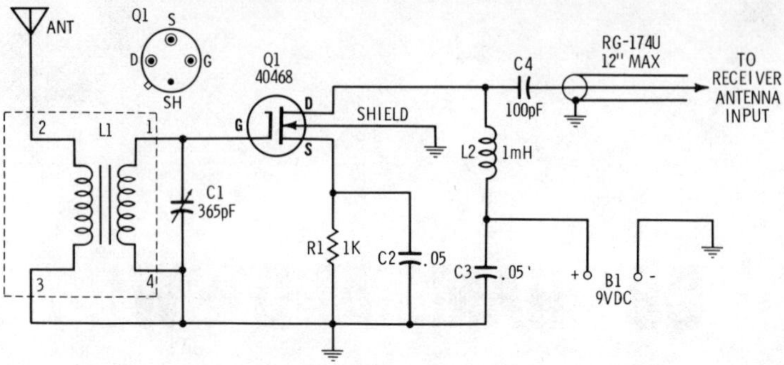

Fig. 21A. Be extremely careful when installing or removing the insulated-gate field-effect transistor (IGFET) or the metal oxide semiconductor field effect transistor (MOSFET).

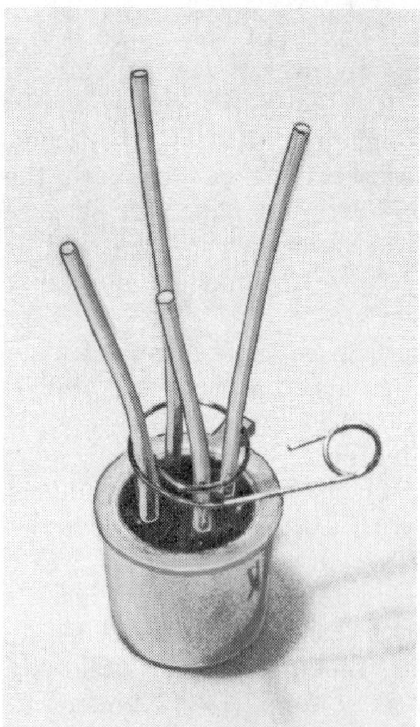

Fig. 21B. The 40468 transistor is a MOSFET and is supplied with a shorting clip around the four leads. To prevent static charges from a soldering iron, pliers, cutters, or even your fingers, from destroying the transistor, the shorting clip must be retained throughout construction. Remove the clip only after all construction is completed. If you have to remove any insulated-gate type FET from a circuit, be sure to short the leads together with a length of bare wire before it is unsoldered or clipped from the circuit.

PARTS LIST

Item	Description
B1	Battery, 9V, Type 2U6 (or equiv)
C1	Capacitor tuning, 365-pF
C2, C3	Capacitor, 0.05 μF, 100-Vdc
C4	Capacitor, 100-pF, 100-Vdc
L1	Broadcast band antenna coil, Miller Type A-5495A (or equiv)
L2	Choke, 1-mH
R1	Resistor, 1000-ohm, ½-watt
Q1	Transistor, FET, RCA 40468 (or equiv)

22 CB TRANSMISSION LINE MONITOR

This monitor "steals" an insignificant amount of power output from the transceiver, but it keeps a constant watch on the output of a CB rig. If something causes the output to fall off, the monitor instantly lets you know that something is wrong.

The monitor can be built in a small metal cabinet or customized into the transceiver cabinet. Make the wiring between D1, R1, R2 and C1 as short as possible. The pick-up loop consists of four or five turns of insulated, solid, hook-up wire wrapped around an exposed part of the inner conductor of the output coaxial cable.

An alternate pickup is about 6 inches of wire slipped under the coaxial cable shield. The continuity of the shield however, must not be broken.

Vary the number of turns in the loop or the length of the wire to secure an approximate half-scale meter indication with potentiometer R2 set to about center position. The potentiometer serves as a coarse sensitivity control.

PARTS LIST

Item	Description
C1	Capacitor, 0.001-μF, 100-Vdc
D1	Diode, 1N60
M1	Meter, 0-1 mA dc
R1	Resistor, 1500-ohm, ½-watt
R2	Resistor, 10,000-ohm, ½-watt

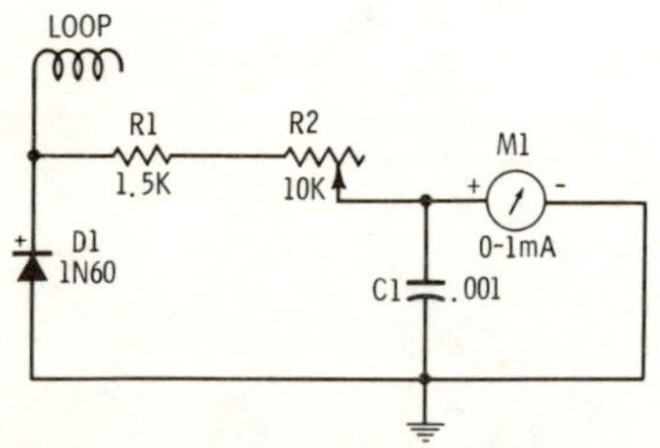

Fig. 22A. This meter lets you know how well your signal is getting out.

Fig. 22B. For mobile use, build the transmission line monitor in the smallest possible cabinet.

23 CB MODULATION METER

You can measure the percent modulation of a CB transceiver with almost the same accuracy as the local broadcast station—because you will be using the same type of peak measuring system.

Meter M1 must be a high-speed model, such as the Alco P-100 series. Keep the leads of R1, D1, and R2 as short as possible. Connect the meter across the transceiver coaxial cable output line by using a "T" connector at the transceiver output jack.

Set switch S1 to *calibrate,* key the transmitter, and adjust R2 for a full scale meter indication. When S1 springs back to the *modulation* position, the meter will indicate the percent modulation when you speak.

PARTS LIST

Item	Description
C1	Capacitor, 500-pF, 100-Vdc
C2	Capacitor, electrolytic, 10-μF, 10-Vdc.
C3	Capacitor, 200-pF, 100-Vdc
C4	Capacitor, 300-pF, 100-Vdc
D1, D2, D3	Diode, 1N60
M1	Meter, high-speed, 0-1 mA dc
R1	Potentiometer, 1000-ohm
R2	Resistor, 560-ohm, ½-watt
R3	Resistor, 910-ohm, ½-watt
S1	Switch, spring-return spdt

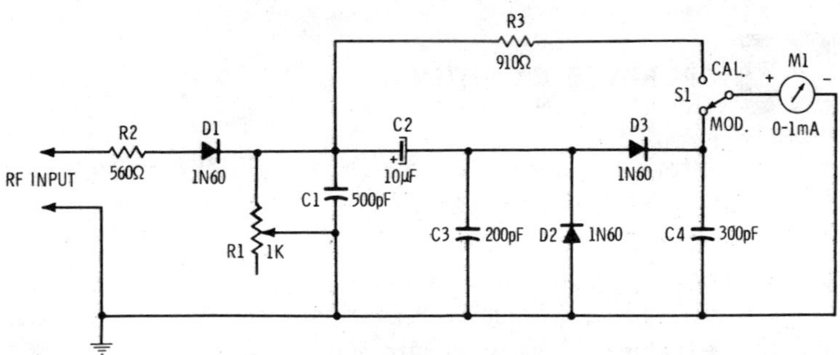

Fig. 23. You can keep a check on the percent of modulation with this meter.

24 A-M MODULATION MONITOR

This simple modulation monitor for a-m ham transmitters requires no connection to the transmitter. Just position the loop near the final tank or antenna matching coil until the signal is heard in the headphones.

PARTS LIST

Item	Description
C1	Capacitor, 100-pF, 100-Vdc
D1	Diode, 1N60
L1	Coil, 3 turns on 1 ½-inch diameter form. Any fine, insulated wire.
Misc.	Headphone, magnetic 2000 ohms or higher

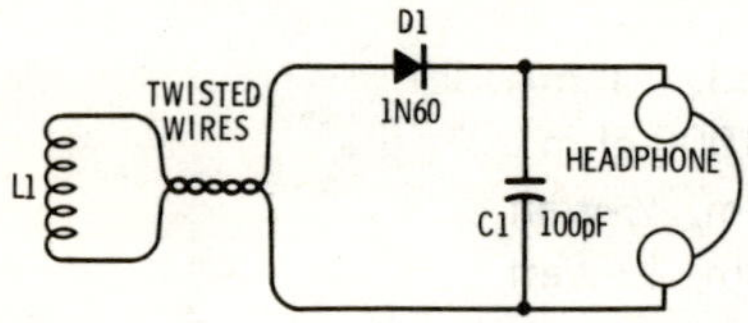

Fig. 24. Here is an economical way to monitor a ham transmitter.

25 PORTABLE CB ANTENNA

A large antenna is always better than a small one, so why use a dinky loaded whip for portable work? You can make a *roll-up*, full-size coaxial antenna from a length of RG-59U coaxial cable.

Cut away the outer insulation for a length of 108 inches and fold the shield braid back over the insulated cable as shown in Fig. 25. Attach a glass or ceramic insulator to the end of the

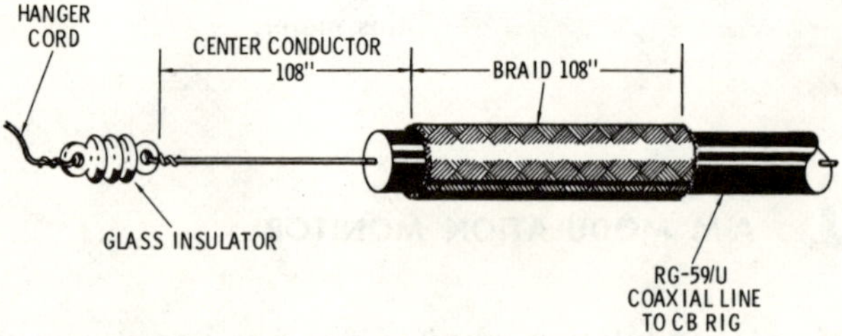

Fig. 25. Here is an antenna you can roll up and take with you.

center conductor and hang the antenna from a tree, roof, pole, or window. Attach the lower end of the cable to your transceiver. Keep the antenna as far as possible from metal poles, gutters, and buildings.

26 EXTENDED CB MOBILE WHIP

The average 108-inch CB bumper or fender-mounted whip has a radiation resistance nowhere near the desired 52 ohms. An improvement in the transmission line match, a lower angle of radiation, and more gain are obtained by using an 11½-ft. extended whip antenna. The extended whip antenna however, is not resonant on the CB band, and it must be *electrically* short-

PARTS LIST

Item	Description
C1	Capacitor, variable 100-pF Hammarlund Type MAPC (or equiv)
Misc.	Extension for standard whip antenna

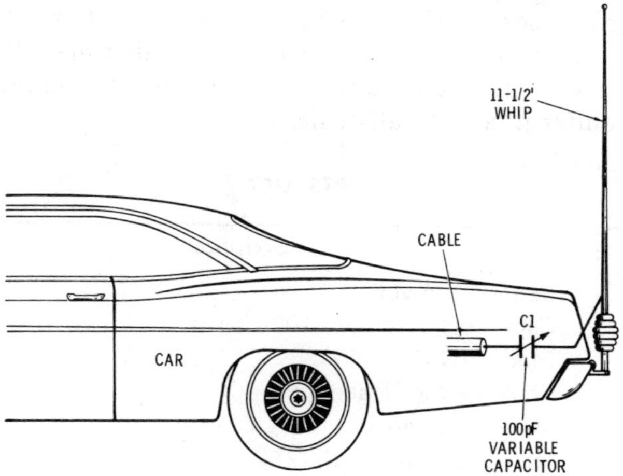

Fig. 26. This extended whip will give you that added gain.

ened to resonance. This is done by connecting a small variable capacitor between the bottom of the antenna and the center conductor of the coaxial cable. Adjust the capacitor for the lowest standing-wave ratio (SWR).

27 SUPERSENSITIVE FIELD-STRENGTH METER

A kilowatt transmitter might pin the needle of regular field-strength meters (FSM) but you need extra high sensitivity to get useful meter indications from low-power oscillators, flea-power transmitters, and CB walkie-talkies.

This simple, easy-to-build, amplified FSM has a sensitivity 150 to 300 times that of ordinary field strength meters. It indicates full scale when other meters cannot budge off the pin. The dependable frequency range is approximately 3 to 30 MHz. A metal enclosure, with a stiff wire antenna about six inches long, is recommended. For compactness, RFC should be a miniature 2.5-mH choke.

To operate, adjust R1 for a ⅓- to ¾-scale indication. Avoid working too close to the top of the scale, since Q1 might saturate and produce full-scale indications at all times. Back off on R1 as you make transmitter adjustments to maintain the meter pointer at about half-scale.

PARTS LIST

Item	Description
B1	Battery, 1.5-volt
C1	Capacitor, 0.001-μF, 100-Vdc
D1	Diode, 1N60
L1	Choke, 2.5-mH (see text)
M1	Meter, 0-1 mA dc
Q1	Transistor, npn, 2N3391
R1	Potentiometer, 50,000-ohm

Fig. 27. Here is one way to measure those weak signals.

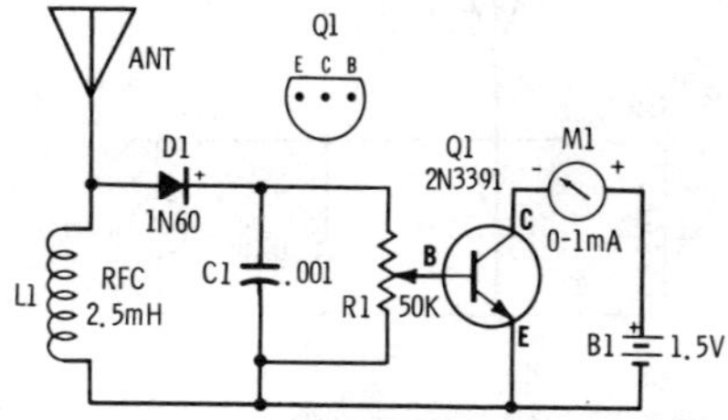

28 TUNABLE FSM

Reasonably high sensitivity without using amplification is obtained when an FSM is tuned to the operating frequency. If a *poly*-type miniature capacitor is used for C1, the FSM can be built in a pocket-size cabinet.

The tuning range is from 1.5 to 144 MHz, depending on the choice of coil L1. A series of coils can be made with phone tips, and plug-in connections can be used on the chassis to make it easy to tune different bands. Consult any coil table to get winding data for the specific frequency coverage. An example of a coil is shown across the variable capacitor in the photograph of Fig. 28B.

Greater sensitivity is obtained if a more sensitive meter is used. A 50-μA meter provides maximum sensitivity with reasonably rugged construction.

PARTS LIST

Item	Description
C1	Capacitor, 365-pF variable
C2	Capacitor, 0.005-μF, 100-Vdc
D1	Diode, 1N60
J1, J2	Phone tip and jack
L1	Coil (see text)
M1	Meter, 0-1 mA dc (see text)

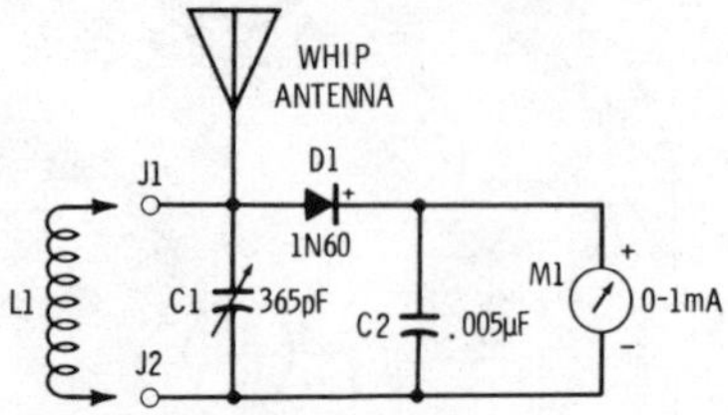

Fig. 28A. Tune in different bands by making coil L1 a series of plug-in coils.

Fig. 28B. For optimum sensitivity on 6 and 2 meters (50 and 144 MHz) use an APC or MAPC type variable capacitor. Also, keep the coil, capacitor, and diode as close together as possible as shown here.

29 HEADPHONE NOISE AND VOLUME LIMITER

Most receivers, particularly the budget variety, do not provide automatic volume control for code reception. Thus, a continuous wave signal that blows your headphones off one moment might lie buried on the threshold of hearing the next. The headphone limiter chops those "Hundred over S9" signals down to size so they equalize with weaker signals, giving relatively constant headphone volume. It also chops noise pulses down to tolerable levels.

Because the clipping action produces some distortion, the limiter should feed a headphone Q-Peaker described in the following project. The value of resistor R2 should match the existing speaker impedance and power; in most instances, this will be 4 ohms at 2-to-5 watts.

PARTS LIST

Item	Description
D1, D2	Diode, 1N60
R1	Potentiometer, audio taper, 5000-ohm
R2	See text

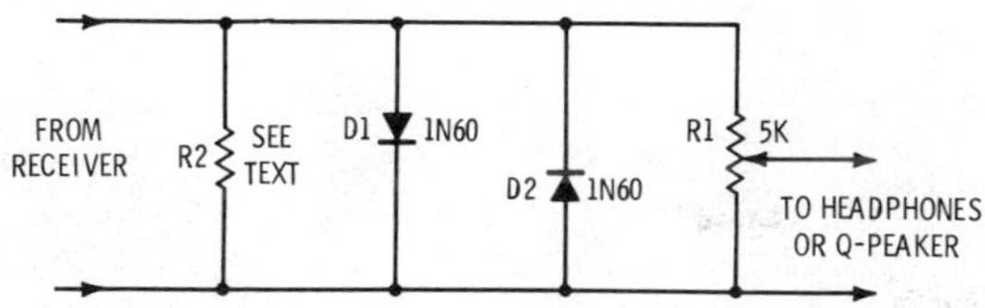

Fig. 29. Eliminate those loud bursts of sound and noise from the headphones.

30 HEADPHONE Q-PEAKER

If you are tired of copying cw signals through the noise, the 29 cent Q-peaker is the next best thing to a Q-multiplier—and it's a lot less expensive.

Capacitor C1 plus the inductance of magnetic headphones form a parallel resonant circuit at approximately 1 kHz. All other signal frequencies are sharply attenuated so you hear mainly the desired signal, which has been peaked with the bfo.

Resistor R1 isolates the resonant circuit to prevent the low impedance output of the receiver from reducing the "Q" of the headphone circuit.

The exact value for C1 depends on the particular headphones employed; try different values in the 0.005- to 0.05-μF range until the desired peaking action is obtained.

PARTS LIST

Item	Description
C1	Capacitor, 0.005 to 0.05-μF, 100-Vdc (see text)
R1	Resistor, 100,000-ohm, ½-watt
Misc.	Headphones, magnetic, 1000- to 3000-ohm

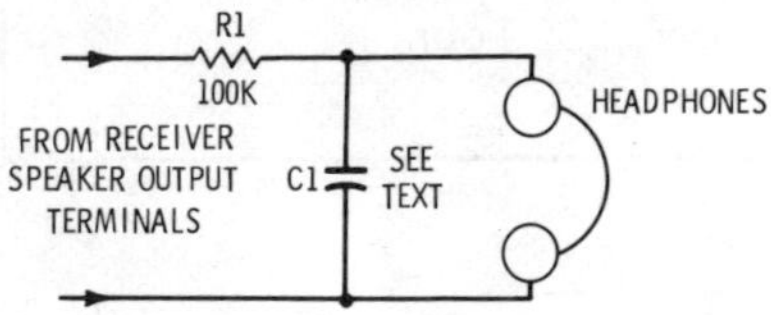

Fig. 30. This Q-peaker is connected in place of the speaker.

31 S-9er FOR SWLs

Super sensitivity is the feature of this short-wave preselector. It provides overall gain as high as 40 dB from 3.5 to 30 MHz.

Diode D1 protects the FET against excessive gate voltage caused by nearby transmitters. Transistor Q2 serves as an emitter-follower to match the medium output impedance of FET Q1 to the low input impedance of the receiver.

Since transistor Q1 is a MOSFET type with a gate that is sensitive to static charges, it must be handled with a short circuit across all leads until the construction is finished. Also, a soldering iron must not be applied to any of the leads on the FET unless they are all shorted together.

RG-174U coaxial cable should be used between the output of this preselector and the input of a receiver to prevent any feedback to the preselector input.

PARTS LIST

Item	Description
C1	365-pF capacitor, tuning
C2, C3	Capacitor, 0.05-μF, 25-Vdc
C4	Capacitor, 500-pF, 100-Vdc
D1	Diode, 1N914
L1	Antenna coil, 1.7-5.5 kHz use Miller B-5495A 5.5-15 MHz use Miller C-5495A, and 12-36 MHz use Miller D-5495A.
Q1	Transistor, FET, RCA 40468 (or equiv)
Q2	Transistor, npn, 2N3394 (or equiv)
R1	Resistor, 470-ohm, ½-watt
R2	Resistor, 2400-ohm, ½-watt
R3	Resistor, 4700-ohm, ½-watt

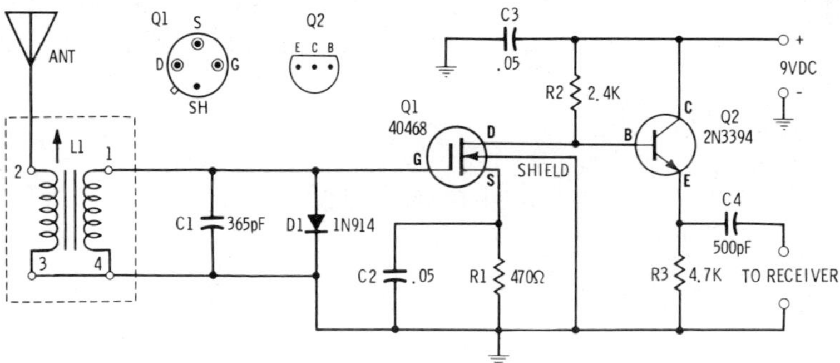

Fig. 31A. This preselector will help you pull in those weak stations.

Fig. 31B. Two switched coils will provide coverage from approximately 3.5 to 30 MHz if the second section of a dual 365-pF capacitor is switched in on the low-band coil. A third switch position can bypass the S-9er, connecting the antenna directly to the receiver. A vernier dial with calibrations will allow the instant preset of the tuned frequency.

32 100-KHZ FREQUENCY STANDARD

Few short-wave receivers below the deluxe class have accurate dial calibration. However, with a 100-kHz frequency standard, you will know with great precision the frequency to which the receiver is tuned.

The frequency standard is a common-base oscillator that produces sufficient radiated signal when it is built in a plastic cabinet. When a metal cabinet is used, a short, 12-in. antenna connected to the collector of Q1 through a 50 μF capacitor is required. In some instances, the antenna may have to be connected to the receiver antenna terminal.

Wiring is not critical, and almost any layout will work. If the oscillator does not start, change the value of R2 in 20 per-

PARTS LIST

Item	Description
B1, B2	Battery, 1.5-volt "AA" or "AAA" cell
C1	Capacitor, 0.01-μF, 25-Vdc
C2	Capacitor, silver mica, 200-pF
L1	Coil, 2 to 18 mH
L2	Choke, RFC, 2.5 mH
Q1	Transistor, npn, Motorola HEP-641 (or equiv)
R1	Resistor, 750,000-ohm, ½-watt
S1	Switch, spst
X1	Crystal, 100-kHz

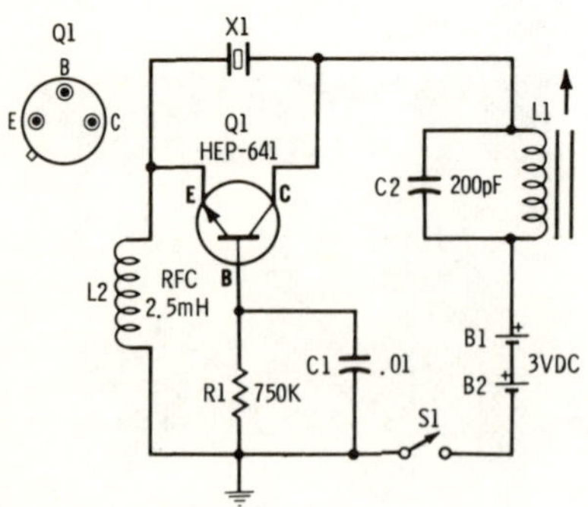

Fig. 32. Use this crystal-controlled oscillator to calibrate your receiver.

cent increments until you get consistent oscillator operation. If you want to calibrate the crystal with WWV, install a 50-pF trimmer in series or in parallel with the crystal.

33 FREQUENCY SPOTTER

You can find weak, short-wave signals if you use this frequency spotter. Obtain crystals cut for frequencies on or near your favorite short-wave stations. Plug them into the spotter, and transmit powerhouse markers on the short-wave bands. If your receiver has a bfo, it will emit a loud "beep" when the spotter is tuned in. If your receiver has no bfo, simply tune around the frequency until the receiver gets deathly quiet. Either way, your receiver gets calibrated.

The spotter can be assembled on a small section of perforated board and flea clips can be used for tie points. For good performance, all components must be firmly mounted and well soldered. A Type 2U6 battery can be used for the power supply.

The crystals for this circuit are the *fundamental* type, rather than the overtone type. Many low-cost surplus crystals are available. Even if you cannot get the exact frequency, 25¢ will buy a crystal close to the exact frequency.

PARTS LIST

Item	Description
B1	Battery, 9V, Type 2U6 (or equiv)
C1	Capacitor, silver mica, 1200-pF
C2	Capacitor, silver mica, 75-pF
C3	Capacitor, 250-pF, 100-Vdc
C4	Capacitor, 0.01-μF, 100-Vdc
L1	Choke, 1 mH
Q1	Transistor, pnp, RCA 2N274 (or equiv)
R1	Resistor, 220,000-ohms, ½-watt
R2	Resistor, 1000-ohm, ½-watt
X1	Crystal socket

A direct connection between the spotter and the receiver is not needed. Simply position the spotter near the receiver antenna and start tuning until you hear the markers.

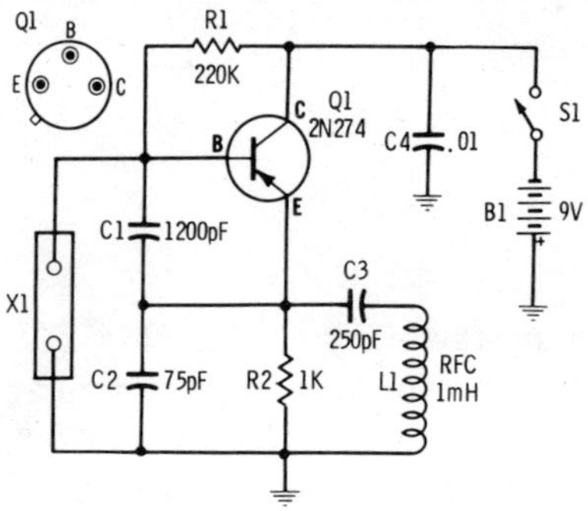

Fig. 33. With the right crystals, this spotter will permit you to set the receiver for any desired frequency.

34 BUDGET CPO

Spare components are often all that is needed for this budget code practice oscillator (CPO). The tone frequency is determined by the value of C1 and C2, and can be changed by simply increasing or decreasing the values. The ratio between the two, however, must be maintained; that is, capacitor C2 should always have about 10 times as much capacity as capacitor C1.

PARTS LIST

Item	Description
B1	Battery, 4.5-volt, three "AA" or "AAA" cells
C1	Capacitor, 0.02-μF, 25-Vdc
C2	Capacitor, 0.22-μF, 25-Vdc
Q1	Transistor, npn, Motorola HEP-641 (or equiv)
R1	Potentiometer, 50,000-ohm, ½-watt
R2	Resistor, 1500-ohm, ½-watt
R3	Resistor, 27,000-ohm, ½-watt
R4	Resistor, 2700-ohm, ½-watt
Misc.	Magnetic headphones

The component values given in Fig. 34 produce a tone frequency of approximately 800 Hz. The battery current drain is about 1 mA and will provide many hours of code practice.

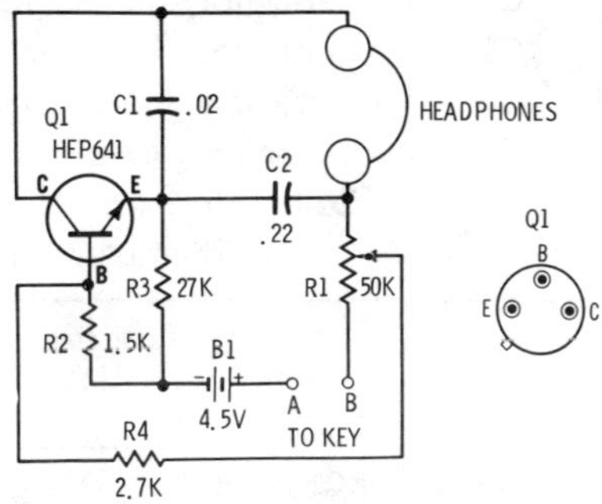

Fig. 34. This code practice oscillator will produce a tone of about 800 Hz.

35 CB TUNING ADAPTER

A crystal-controlled CB receiver that uses overtone-type crystals and has an i-f of 1300 to 1500 Hz can be converted to full 23-channel coverage with this tuning adapter. It works only on circuits where the crystal connects from the grid of the oscillator tube to ground.

The form for coil L1 is a 1-inch diameter wood dowel. Wind the coil on the form as tightly as possible and stretch it to a length of one inch. Make the connection to the transceiver with the shortest possible lengh of RG-58A/U coaxial cable. The shield is connected to the transceiver chassis and to the bottom end of coil L1.

Set tuning capacitor C3 so that its plates are fully meshed (closed), and then adjust trimmer C2 until channel 1 is received. Depending on the i-f of the receiver, C3 might tune slightly above or below the frequency of the Citizens Band. If so, change the value of C1 very slightly to obtain only 23-channel coverage. If capacitor C1 is made smaller, the tuning range will be reduced.

PARTS LIST

Item	Description
C1	Capacitor, silver mica, 10-pF
C2	Capacitor, trimmer, 7.5-pF
C3	Capacitor, miniature variable, 30-pF
C4	Capacitor, 47-pF, 100-Vdc
L1	Coil, 5 turns of No. 16 enameled wire wound on a 1-in. diameter form spaced 1 in. to end.

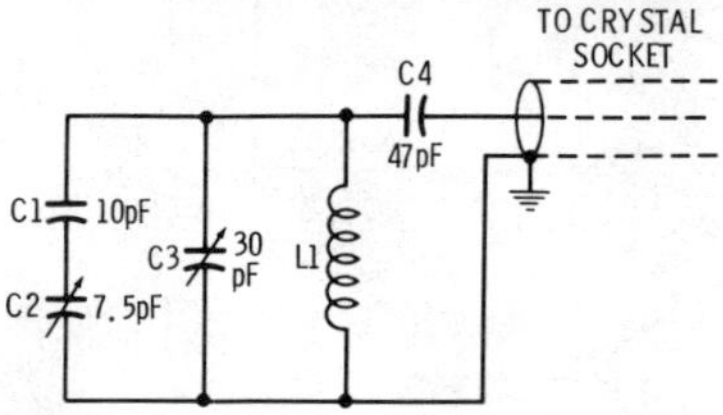

Fig. 35. This circuit can be used as a replacement for a crystal to permit tuning of the local oscillator.

36 ELECTRONIC KEYER

This is not the equal of a $50 electronic keyer, but it is a lot easier to use than a hand key.

When the paddle terminal of the key makes contact with the dot terminal, C1 starts to charge. When the voltage of C1 causes Q1 to conduct, the collector current pulls in relay K1, thereby closing contacts A and B and keying the transmitter. When relay contacts B and C open the paddle circuit, C1 starts to discharge through resistors R1, R2, and R5, transistor Q1 stops conducting and relay K1 drops out. This causes C1 to recharge and the cycle is repeated. The dashes work in a similar fashion.

Potentiometer R1 sets the dot-dash ratio, and potentiometer R2 sets the speed. Potentiometer R3 controls the drop-out of relay K1. The relay will drop out just before transistor Q1 stops conducting. Adjusting potentiometer R3 has a slight effect on the dot-to-space ratio.

PARTS LIST

Item	Description
C1	Capacitor, electrolytic, 3-μF, 6-Vdc
C2	Capacitor, electrolytic, 10-μF, 6-Vdc
D1	Diode, 1N60
K1	Relay, 12-Vdc, P & B RS-5D-12 (or equiv)
Q1	Transistor, pnp, Motorola HEP-253 (or equiv)
R1	Potentiometer, linear, 10,000-ohm
R2	Potentiometer, 50,000-ohm
R3	Potentiometer, 1000-ohm
R4	Resistor, 560-ohm, ½-watt
R5	Resistor, 1200-ohm, ½-watt

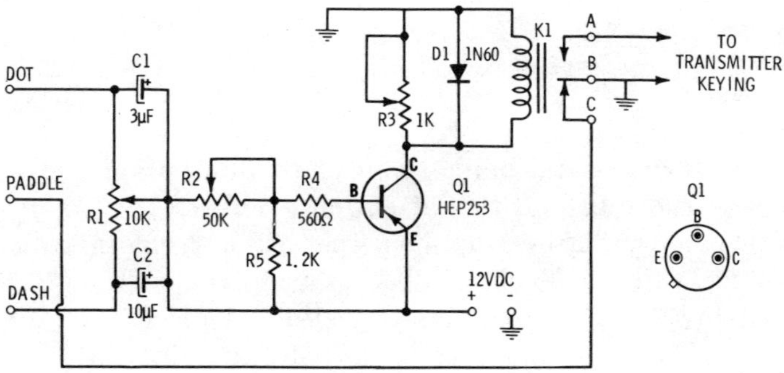

Fig. 36. This electronic keyer is designed to work with a semiautomatic key that is usually referred to as a "bug."

EXPERIMENTAL CIRCUITS

37 LIGHT FLASHER

If a light winks and blinks, someone will stop and look—that is the purpose behind this attention grabber.

When power is first applied, there is current in transistor Q2, and I1 lights. Then, feedback through capacitor C2 causes transistor Q1 to conduct. As capacitor C1 discharges through resistor R4, transistor Q2 is cut off, thereby extinguishing lamp I1. When the voltage of capacitor C1 equalizes, Q2 again turns on and the cycle is repeated.

PARTS LIST

Item	Description
C1	Capacitor, electrolytic, 10-μF, 15-Vdc
C2	Capacitor, electrolytic, 30-μF, 15-Vdv
C3	Capacitor, 0.2-μF, 25-Vdc
I1	Lamp, No. 49
Q1, Q2	Transistor, npn, 2N3394
R1	Potentiometer, 1-megohm
R2	Resistor, 4700-ohm, ½-watt
R3, R4	Resistor, 10,000-ohm, ½-watt
R5	Resistor, 120-ohm, ½-watt

Potentiometer R1 determines the turn-on/turn-off rate of the transistors and, hence, determines the rate at which the light will blink.

The npn transistors can be replaced by pnp types if the polarity of the battery and capacitors C1 and C2 are reversed.

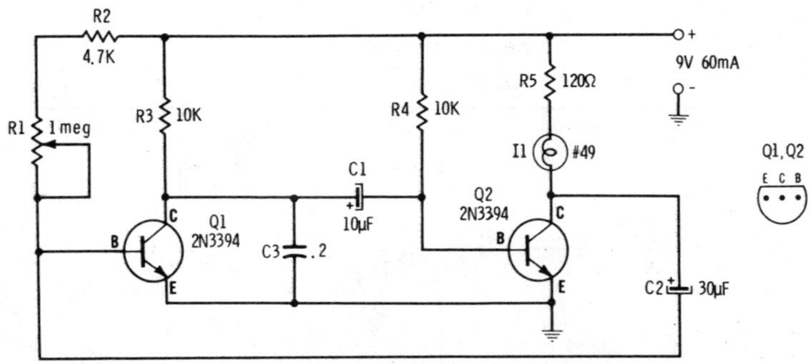

Fig. 37. This is a flip-flop multivibrator which lights a lamp when transistor Q2 is in the conducting state.

38 LIGHT COMPARATOR

The light comparator will check or make it possible to adjust two light sources for equal intensity. The metering circuit is a balanced bridge consisting of R1, R2, R3, Q1, and Q2. With solar cells R1 and R2 exposed to the same light source, balance control R1 is adjusted for a zero meter indication.

In operation, unequal light falling on solar cells R1 and R2 changes the base bias of the transistors. This upsets the balance of the collector currents and causes the meter to indicate.

PARTS LIST

Item	Description
B1	Battery, 9-volt Type 2U6 (or equiv)
M1	Meter, 0-1 mA dc, zero center
PC1, PC2	Solar cell
Q1, Q2	Transistor, pnp, 2N109
R1	Potentiometer, 5000-ohm
R2, R3	Resistor, 1000-ohm, ½-watt

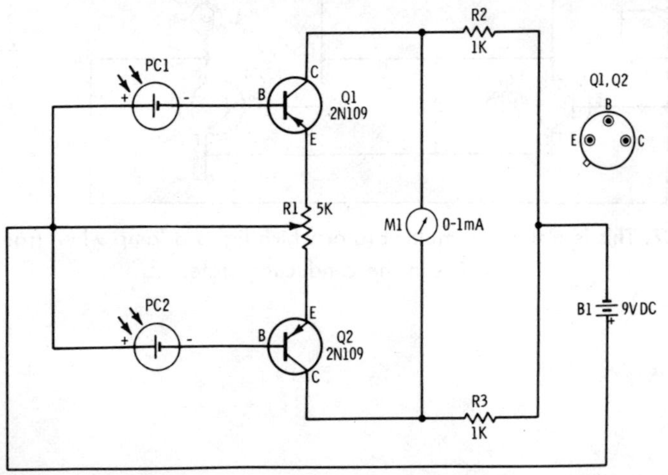

Fig. 38. This circuit makes it an easy job to compare the intensity of two light sources.

39 THREE-WAY TONE GENERATOR

Add a terminal or two, and an ordinary code practice oscillator (CPO) will also serve as a tone generator, or an intruder alarm.

The circuit shown in Fig. 39 is a Hartley oscillator whose tone frequency is determined by the value of resistor R2. Just about any layout will work, but transformer T1 must be an output transformer of the type used in table model radios. A miniature transistor transformer might not oscillate—if it does, it will produce raucous tones.

For CPO operation, connect a hand key across terminals C and D. For a "make" intruder alarm connect one or more normally open detectors across terminals C and D. For a "break" intruder alarm, connect a jumper across terminals A and B, which disables the oscillator although power is applied. An intruder breaking a series circuit, or a normally closed detector, will cause the alarm to sound off.

For use as a signal generator, connect a shielded test lead across the speaker terminals.

PARTS LIST

Item	Description
B1	Battery, 9-volt, Type 2U6 or larger
C1, C2	Capacitor, 0.02-μF, 25-Vdc
Q1	Transistor, npn, 2N3394
R1	Potentiometer, 250,000-ohm
R2	Resistor, 10,000-ohm, ½-watt
SP1	Speaker, 3.2-ohm
T1	Output transformer, 5000-ohm center-tapped primary, 3.2-ohm secondary (must **not** be a miniature transistor type)

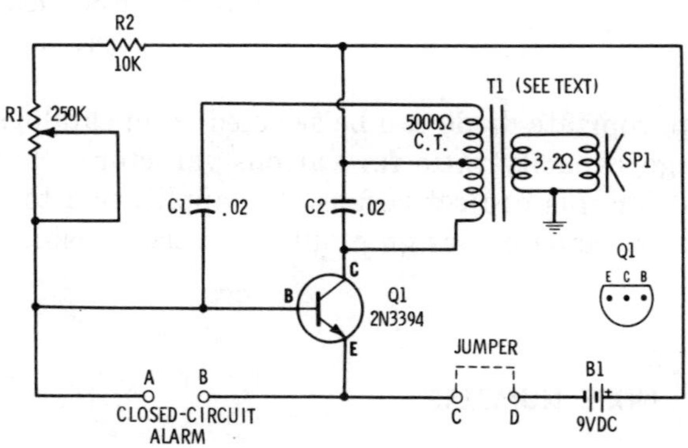

Fig. 39. This oscillator can be turned on by either a "make" or a "break" circuit.

40 ELECTRONIC TUNING

We cannot assign specific values, because each situation is different. However, Fig. 40 shows a basic circuit in which a *varactor* is used for electronic tuning. The varactor is a diode in which the capacitance between the anode and the cathode is determined by the applied voltage. If a varactor is substituted for a tuning capacitor in an LC resonant circuit, the tuned frequency is determined by the value of the applied voltage.

Inductor L1 and the varactor diode form an LC parallel resonant circuit. The dc blocking capacitor prevents the power supply from shorting to ground, and should have nearly zero impedance at the tuned frequency. The capacitance of the blocking capacitor should be at least 20 times the maximum capacitance of the varactor. In effect, the blocking capacitor is a short circuit at the frequency of the tuned circuit and the frequency is controlled by the values of inductor L1 and the varactor diode.

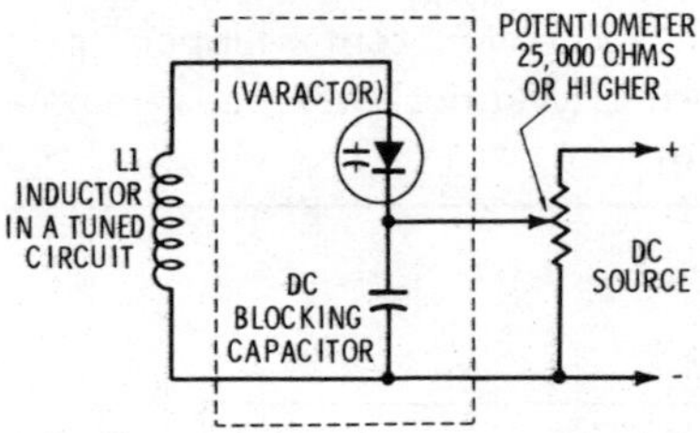

Fig. 40. The varactor control circuit (shown in the dotted lines) can be substituted for the tuning capacitor in an rf oscillator. The dc control source can be located at a distance to provide remote tuning of the oscillator.

The appropriate diode can be selected from catalogs that list the voltage/capacity ratio for various varactors. No parts list is offered for this project because the application to a particular tuner or oscillator design requires special consideration.

41 NIXIE NUMBERS

Using *Nixie* tubes, you can transmit numerical signals or even ball scores over long distances.

The Nixie—a peanut-size tube—has ten numerically shaped neon lamps (0 thru 9). By shorting the appropriate lead to ground, an internal neon lamp corresponding to that number is illuminated. Transformer T1 is a center-tapped (CT) 250-volt transformer that provides a peak dc output voltage of approximately 175 volts. Although current requirements are very low, diodes D1 and D2 should be line-voltage type silicon rectifiers of 200 mA minimum.

PARTS LIST

Item	Description
C1	Capacitor, electrolytic, 30-μF, 250-Vdc
D1, D2	Rectifier, silicon, 200-mA, 400-PIV
T1	Transformer, 120-volt primary, 250-volt, 25-mA center-tapped secondary
V1	Neon readout tube tube (Nixie), National Electronics NL840 (or equiv)

The same power supply can be used for additional Nixie tubes. Each additional Nixie tube connects to the top of capacitor C1.

The neon numbers can be turned on either through an 11-position (one Position for off) rotary switch or individual toggle switches.

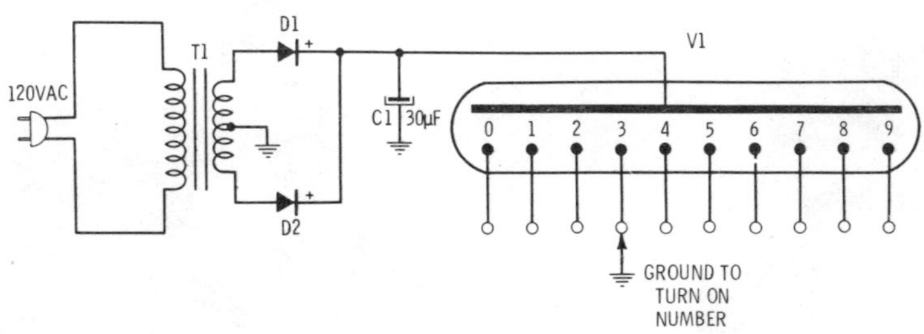

Fig. 41. The numbers in a *Nixie* tube can be lighted with this simple power supply.

42 SIDEBAND SCRAMBLER

Feed audio modulation to one input, a carrier to the other input, and the output of the sideband generator will be the upper and lower sidebands with a suppressed carrier. Where is it used? Try a sideband transmitter-receiver or a telephone speech scrambler.

PARTS LIST

Item	Description
D1, D2, D3, D4	Diodes, 1N60

To recover the original speech connect the scrambled signal into the modulation input. The carrier input must be the same frequency as was used to scramble the speech.

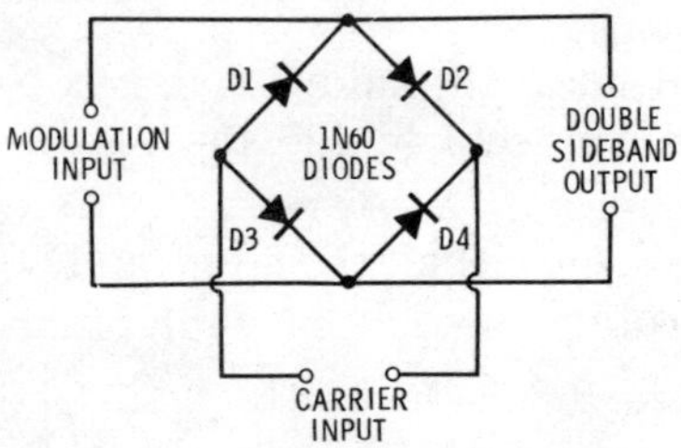

Fig. 42. For best results, select the four diodes so that they have similar conduction characteristics.

LAMP AND MOTOR CONTROL CIRCUITS

43 400-WATT LAMP DIMMER

With miniature components and extreme care, you can build a low-power lamp dimmer right inside a lamp socket. Without a heat sink, triac Q1 can control up to a 400-watt load.

PARTS LIST

Item	Description
C1, C2	Capacitor, 0.068-μF, 200-Vdc
I1	Lamp, neon, NE-83 or NE-2
I2	External lamp not to exceed 400 watts
Q1	Triac, RCA 40502 (or equiv)
R1	Potentiometer, 50,000-ohm, 2-watt
R2	Resistor, 15,000-ohm, ½-watt

Instead of a relatively expensive trigger diode, an ordinary neon lamp of the NE-83 or NE-2 variety is used. (An NE-83 is treated for dark operation, and will provide more consistent performance.) Because the neon lamp does not trip the triac gate until it conducts, the lamp turns on at medium brilliance. It can then be backed down to a soft glow.

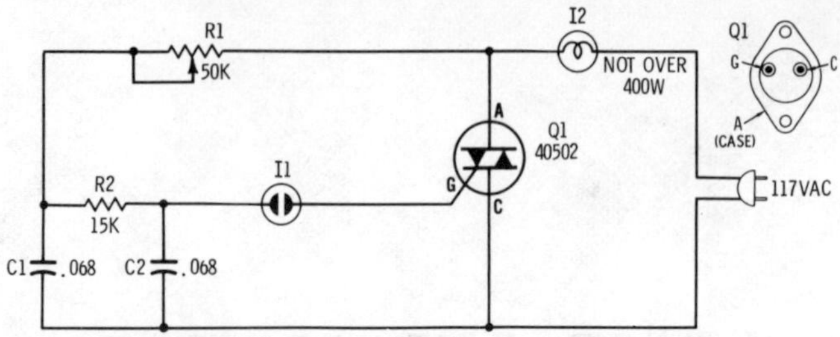

Fig. 43A. This low cost light dimmer will control 400 watts.

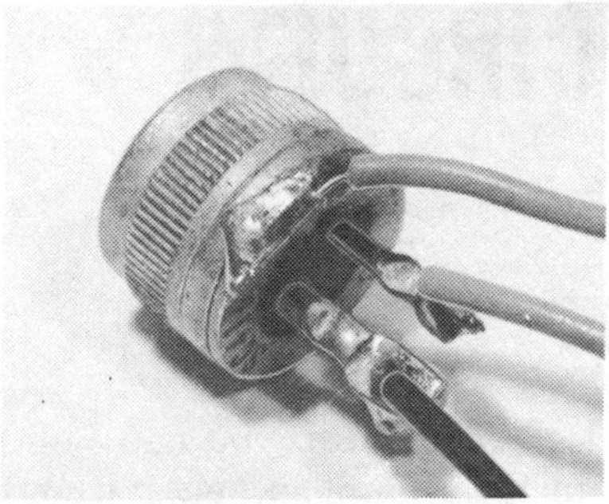

Fig. 43B. Press-fit triacs that use the case as a cathode or anode terminal are often available on the surplus market at rock-bottom prices. They can be easily substituted for the three terminal types by soldering a connecting lead directly to the case. Use a soldering iron rated between 100 and 150 watts and a nonacid soldering paste (flux).

44 MINIDRAIN PILOT LAMP

If you need a pilot lamp for portable equipment that will not burn up the batteries in minutes, then try *high frequencies* for the answer. This is how the circuit in Fig. 44 works: Transistor Q1 serves as a blocking oscillator with the frequency determined by resistors R1 and R2, and capacitor C1. As transistor Q1 oscillates at an audio frequency, current in the collector circuit of Q1 and in the transformer secondary induces a high voltage in the primary of transformer T1. This becomes the high-voltage, low-current source to light neon lamp I1.

Adjust potentiometer R2 so the frequency is high enough to keep lamp I1 constantly illuminated. If you want a warning device, potentiometer R2 can be adjusted so the neon lamp blinks on and off at a rapid rate.

PARTS LIST

Item	Description
C1	Capacitor, electrolytic, 3-μF, 25-Vdc
I1	Lamp, neon, NE-2
Q1	Transistor, pnp, Motorola HEP-253 (or equiv)
R1	Resistor, 100,000-ohm, ½-watt
R2	Resistor, 250,000-ohm, ½-watt
T1	Miniature transformer, 5000-ohm center-tapped primary, 8- and 16-ohm secondaries

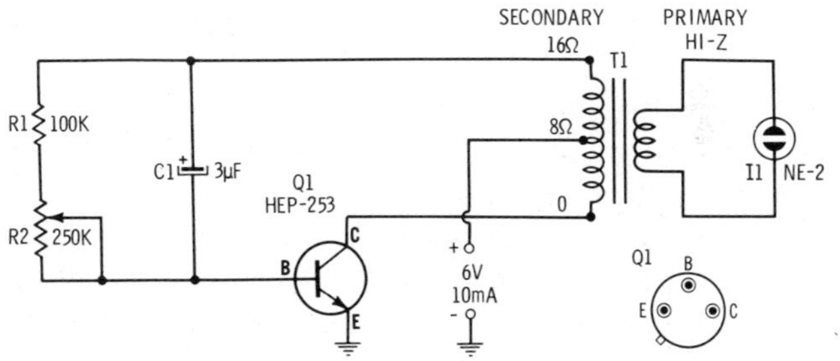

Fig. 44. Circuit for lighting a neon pilot lamp from a 6-volt source.

45 LIGHT-SENSITIVE RELAY

The light-sensitive relay can be adjusted to trip on either a substantial change in illumination, such as night to day, or by a subtle change in illumination, such as an increase or decrease in ambient light level.

Transistor Q1 can be any general purpose pnp transistor of the 2N109/2N217 type although greater sensitivity is obtained with a high gain transistor such as the 2N2613. Relay K1 is a high-sensitivity type such as the Sigma relays used by model control hobbyists.

Potentiometer R2, is part of the voltage divider formed by photocell PC1 and resistors R1 and R2. Potentiometer R2 is adjusted with normal ambient light falling on the photo cell;

PARTS LIST

Item	Description
B1	Battery, 6-volt, Type Z4 (or equiv)
K1	Sensitive relay, 1000-ohm, 2- to 3-mA
PC1	Photocell, RCA 4425 (or equiv)
Q1	Transistor, pnp, 2N2613 (see text)
R1	Resistor, 120-ohm, ½-watt
R2	Potentiometer, 5000-ohm
S1	Switch, spst

the base bias current (through PC1) is just below the value needed to cause the collector current to activate relay K1. When additional light falls on PC1, the photocell resistance decreases, thereby increasing the base bias, which causes the collector current to activate relay K1.

This circuit can be controlled by sunlight so that relay K1 drops out at dusk to turn on a night light. Also a flashlight can be used to trip relay K1 for "killing" television commercials by shorting (or lifting) the speaker connections.

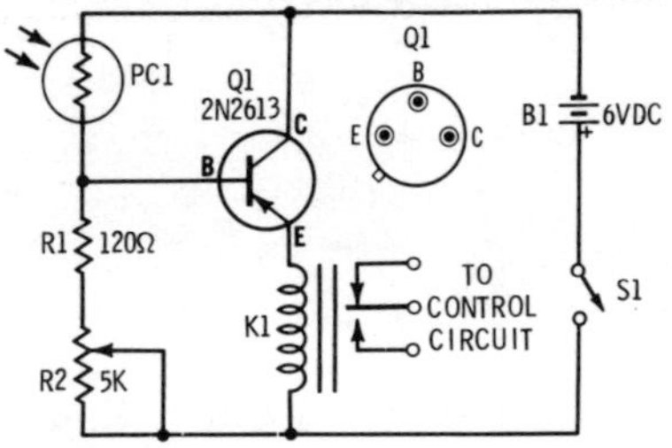

Fig. 45. This circuit can be used to open or close a circuit when light on the photocell is increased or decreased.

46 INDUCTION MOTOR SPEED CONTROL

A triac and four other components are all you need to convert old appliance and shaded-pole motors (salvaged from old phonographs) into slow-speed hobby drills, chemical stirrers, variable-speed turntables, display drives, etc. The reason you can control induction motors with a triac controller is that un-

PARTS LIST

Item	Description
C1, C2	Capacitor, 0.1-μF, 200-Vdc
Q1	Triac, RCA 40431 (with built-in trigger diode)
R1	Potentiometer, 100,000-ohm linear taper
R2	Resistor, 10,000-ohm, 1-watt

like the SCR, which is a rectifier, the triac passes both the negative and positive half-cycles; hence, it can be used to control induction motors.

Unlike other speed control devices, which require an external trigger device such as a neon lamp or a diac, the triac Q1 used in this project combines both the triac and the diac trigger diode in the same case.

The motor used must be limited to a maximum of 6 amperes (or 740 watts). Triac Q1 must be provided with a heat sink, which can be the metal cabinet. Build up a marble-size mound of epoxy adhesive on the cabinet and press the case of Q1 into the epoxy. Make certain that the triac case is not shorted to the cabinet, and that it is insulated by the epoxy.

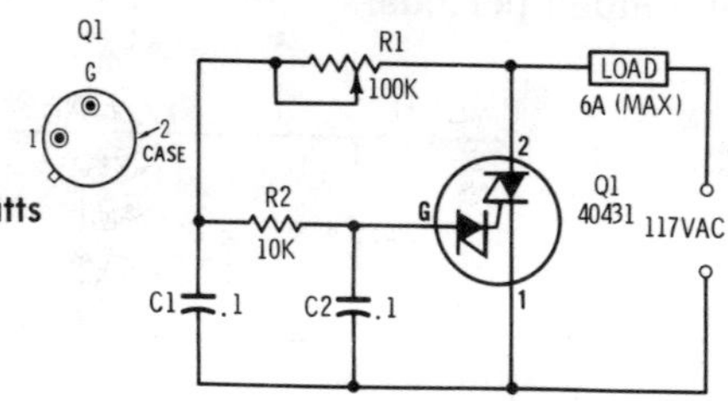

Fig. 46. Loads of up to 140 watts can be controlled by this unit.

47 HIGH TORQUE MOTOR SPEED CONTROL

As the speed of an electric drill is decreased by loading, its torque also decreases. But a compensating motor speed control puts the torque back into slow-speed operation.

When the drill slows down, a back voltage developed across the motor, which is in series with the SCR cathode and gate, decreases. The SCR gate voltage (actually the firing angle) increases as the back voltage is reduced. The extra gate voltage causes the SCR to conduct over a larger angle, and more current is supplied to the drill. When the speed of the drill is being reduced by loading, additional current tends to keep the speed constant.

PARTS LIST

Item	Description
D1, D2	Rectifier, silicon, 500-mA, 200-PIV
F1	Fuse, 3-ampere, Type 8AG "high-speed"
R1	Resistor, 2500-ohm, 5-watt
R2	Potentiometer, 250-ohm, 4-watt
R3	Resistor, 33-ohm, ½-watt
Q1	Silicon controlled rectifier, 3-ampere, 200-PIV

The only construction precaution is the use of an extra-heavy SCR heat sink. The SCR should be mounted in a ¼-inch thick piece of aluminum or copper at least 1 inch square. The aluminum should be at least 2 inches square if you use the drill for extended periods.

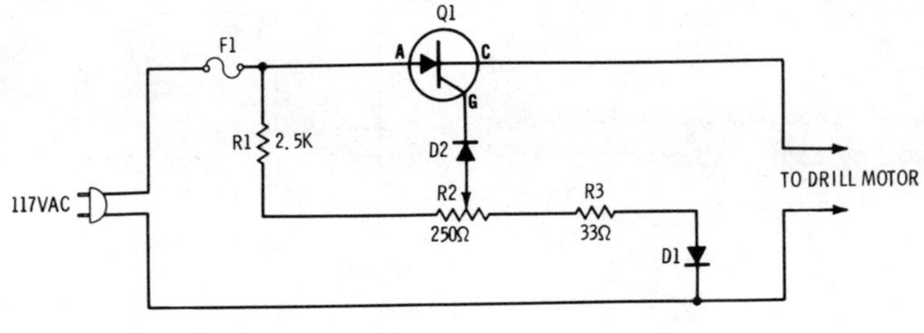

Fig. 47. Circuit for a motor speed control.

48 TRIAC AND SCR HASH FILTER

Triacs and SCRs used by experimenters in light and motor speed controls generate a considerable amount of electrical "hash." This can cause severe interference to broadcast-band

and short-wave radios located within 50 to 100 feet. The noise is generated when ac line current is switched by the triac or SCR. The switching produces sharp pulses of rf energy.

A radio frequency interference filter connected between the triac or SCR and the load will attenuate the interference. The

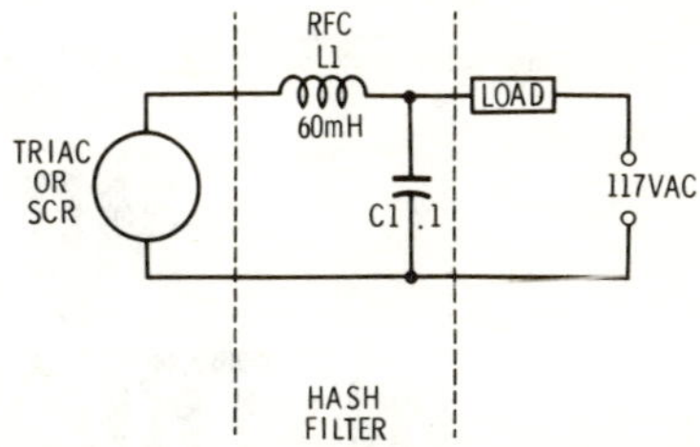

Fig. 48. The rf choke and bypass capacitor should be placed in the same metal box as the SCR or triac control.

best results are obtained if the filter is housed in a metal cabinet with the SCR or triac control. Additional reduction in the interference is accomplished by grounding the cabinet to a cold water pipe or to an electrical ground.

PARTS LIST

Item	Description
C1	Capacitor, 0.1-μF, 200-Vdc
L1	Coil, 60-mH, 65 turns No. 18 AWG magnet wire, 2 layers wound on a 3 × ¼-inch ferrite rod such as used for a-m broadcast antennas.

MUSICAL INSTRUMENTATION

49 98¢ FUZZBOX

If you want guitar fuzz for that *Now* sound, but can't afford a *fuzzbox,* try this project until your guitar picking earns some extra money.

Install the two diodes (shown in the dotted lines of Fig. 49) and the potentiometer across the volume control of the guitar amplifier. The diodes clip the normal sound waveforms, thereby producing the fuzz effect. Potentiometer R1 sets the degree of

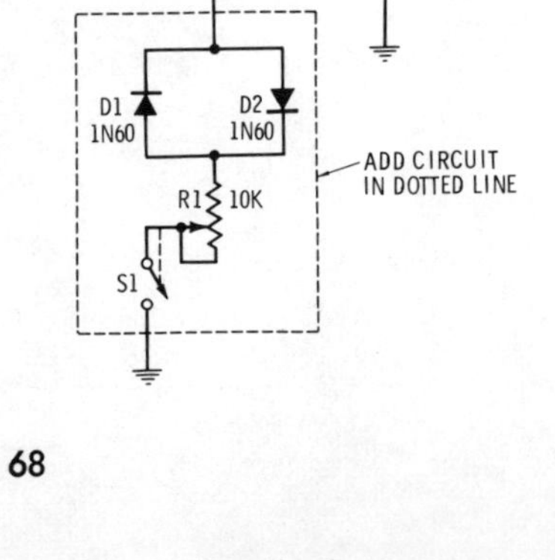

Fig. 49. The circuit shown in the dotted lines is all that you need to add to your present music system.

fuzz; while it is not as much as you would get from a fuzzbox, it *is* fuzz. One restriction on the circuit is that the audio signal level across the volume control must be at least 1-volt—which is generally true in most amplifiers.

PARTS LIST

Item	Description
D1, D2	Diode, 1N60
R1/S1	10,000-ohm miniature potentiometer with switch

50 AMPLIFIED SUPERFUZZ

You can add that real hard-rock sound to any guitar amplifier by connecting the amplified *fuzzbox* shown in Fig. 50 between the guitar and the amplifier. Potentiometer R3 sets the degree of fuzz; potentiometer R8 sets the output level.

Since the fuzz effect cannot be completely eliminated by potentiometer R3, fuzz-free sound requires a bypass switch (shown by the dotted lines) from the input to the output terminals. The bypass switch should completely disconnect the fuzzbox output; the input can remain parallel connected with

PARTS LIST

Item	Description
B1	Battery, 1.5-volt "AA" cell
C1, C3	Capacitor, electrolytic, 0.1-μF, 3-Vdc
C2	Capacitor, electrolytic, 5-μF, 6-Vdc
Q1, Q2	Transistor, pnp, 2N2613
R1	Potentiometer, 1-megohm
R2	Potentiometer, 50,000-ohm, audio taper
R3, R7	Resistor, 22,000-ohm, ½-watt
R4	Resistor, 100,000-ohm, ½-watt
R5, R8	Resistor, 10,000-ohm, ½-watt
R6	Resistor, 18,000-ohm, ½-watt
S1	Switch, spst
S2	Switch, spdt

the bypass. Switches S1 and S2 can be replaced by a single dpdt switch so that when the unit is off, the guitar and amplifier are connected for normal operation.

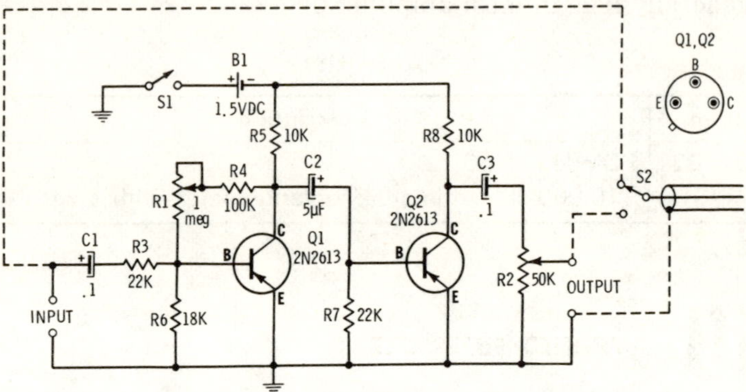

Fig. 50A. Circuit for amplified fuzz sound.

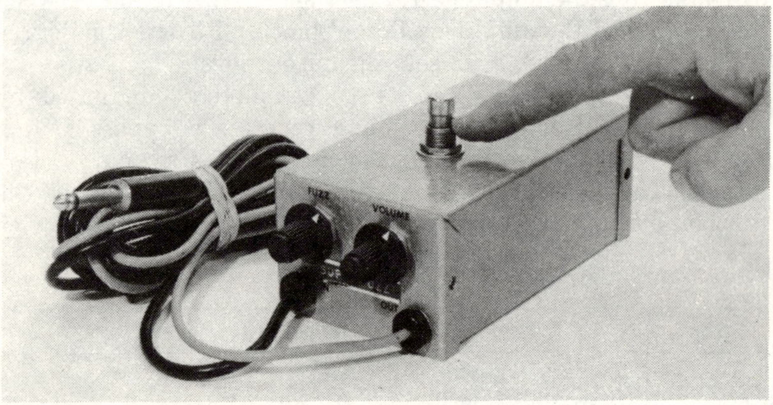

Fig. 50B. If the guitar's signal is bypassed around the Superfuzz with a push-push footswitch the fuzz effect can be keyed in-and-out while playing.

51 METRONOME-TIMER

Providing equally spaced clicks from 3 to 300 per minute, this click generator can be used either as an electronic metronome or as an interval timer, such as used for photo enlarging.

Transistor Q1 functions as an amplifier, but positive feedback from the secondary of transformer T1 to the base of Q1 causes the circuit to regenerate, producing a steady stream of clicks in the speaker. The rate of oscillation, or clicks per minute, is determined by the setting of potentiometer R1 (in commination with resistor R2).

PARTS LIST

Item	Description
B1, B2	Battery, 1.5-volt "AA" cell
C1	Capacitor, electrolytic, 10-μF, 6-Vdc
Q1	Transistor, npn, Motorola HEP-641 (or equiv)
R1	Potentiometer, 1-megohm
R2	Resistor, 7500-ohm, ½-watt
S1	Switch, spst
SP1	Speaker, 3.2-ohm
T1	Miniature transformer, audio, 500-ohm primary, 3.2-ohm secondary

With a little time and patience, a dial affixed to the shaft of potentiometer R1 can be calibrated in *beats per minute* by comparing the clicks against a standard metronome.

If the generator does not click when power is first applied, interchange the two leads from the secondary of T1. It is not necessary to interchange the speaker leads and their connection to C1 and B1—switch only the transformer leads.

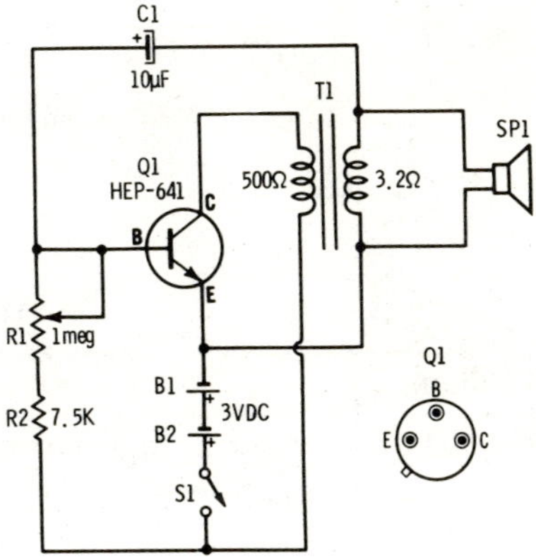

Fig. 51. Circuit for a blocking oscillator that can be used as an audible timer.

PHOTOGRAPHY PROJECTS

52 SLAVE FLASH TRIPPER

Even if you spend $18 or $20 for a professional slave flash tripper, you will get little more than this two-component circuit.

Q1 is a light-activated silicon controlled rectifier (LASCR). The anode-to-cathode circuit is tripped by light entering a small lens built into the top cap.

To operate the slave flash, provide a short length of stiff wire for the anode (A) and cathode (C) connections and terminate the wires in a polarized power plug that matches the sync terminals on your slave electronic flashgun (strobelight). Make certain the anode lead connects to the *positive* sync terminal. Bend the leads so the LASCR lens faces the main flash. When the main flash is fired, the LASCR will fire the slave.

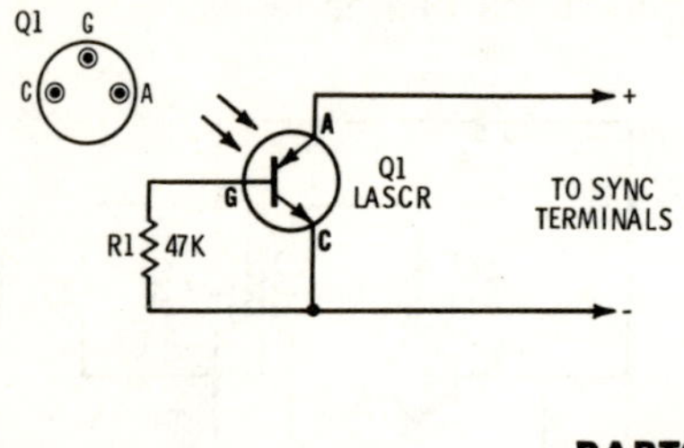

Fig. 52. A light-activated silicon controlled rectifier makes a good trigger for a slave flash.

PARTS LIST

Item	Description
Q1	LASCR (light-activated silicon controlled rectifier), 300-watt
R1	Resistor, 47,000-ohm, ½-watt

No reset switch is needed because voltage at the sync terminals of the flash gun falls below the holding voltage of the LASCR when the flash fires.

53 SLIDE PROJECTOR PROGRAMMER

Soundless slide shows are usually dull. However, a stereo recorder can automate the whole show so that slides change automatically in sync with a prerecorded commentary.

Record your commentary on the left track. At the instant you want slides to change, record a one-second noise or tone burst on the right track. Connect the programmer between the *right* speaker output of the recorder and the remote control cable of the projector. Then make a test run to determine the correct track volume needed to make the noise and/or tone burst activate relay K1. Even a shout can be used as the noise burst.

Start the tape from the beginning. The audience will hear your commentary, music, and sound effects through the speaker connected to the left channel, while the signal on the right channel changes the slides.

PARTS LIST

Item	Description
C1	Capacitor, electrolytic, 25-μF, 50-Vdc
D1, D2	Rectifier, silicon, 500-mA, 100-PIV
K1	Relay, plate-type, 2500-ohm
T1	Transformer, audio output, 5000-ohm center tapped primary, 8-ohm secondary

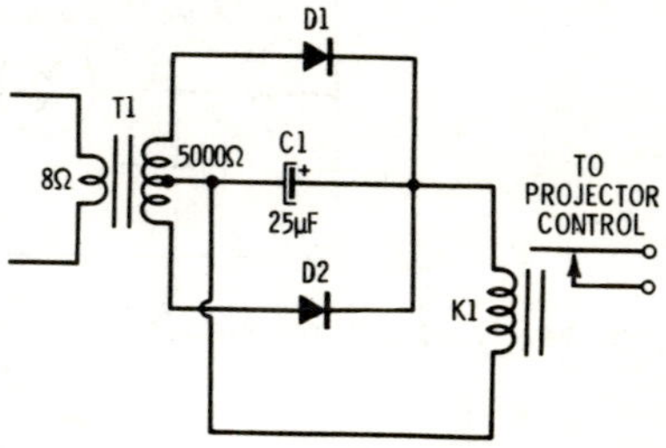

Fig. 53. This circuit can be used to change projector slides automatically.

54 500-WATT PHOTOFLOOD DIMMER

All the flexibility of the variable lighting used in a professional photo studio is yours with this triac controlled lamp dimmer.

Triac Q1 comes supplied with a heat sink which must, in turn, be connected to a larger heat sink. Or you can remove the heat sink of the triac and cement the triac to a metal cabinet. The entire unit is assembled in a metal cabinet with the heat sink of Q1, or just the case of Q1 cemented to the cabinet with epoxy.

Fusing must be employed; otherwise, the surge current when the 500-watt photolamp burns out will instantly destroy Q1. Connect an 8AG (fast-action) 5-ampere fuse in series with the lamp. Do not use 3AG fuses; triac Q1 "blows" faster than a 3AG fuse.

Potentiometer R2 adjusts the lamp intensity from full *off* to essentially 100 percent full *on*.

PARTS LIST

Item	Description
C1, C2	Capacitor, 0.01-μF, 400-Vdc
D1	Diac (trigger diode), RCA 40583 (or equiv)
Q1	Triac, RCA 40638 (or equiv)
R1	Potentiometer, linear taper, 100,000-ohm
R2	Resistor, 1000-ohm, ½-watt
R3	Resistor, 15,000-ohm, ½-watt

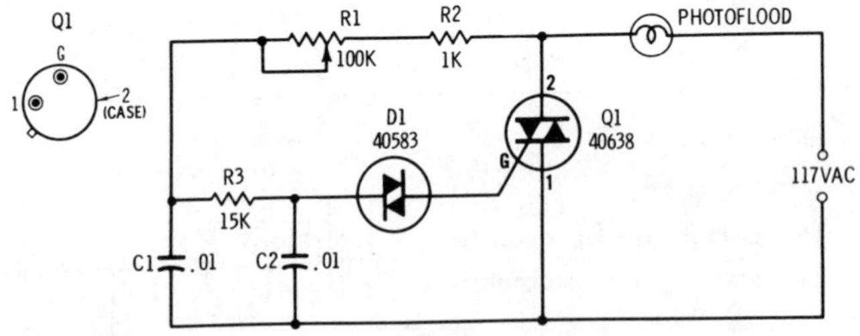

Fig. 54. Triac control circuit for changing the light intensity of a photoflood bulb.

55 ENLARGING METER

The end result of this enlarging meter is that every print is a good print.

When light from the enlarger falls on solar cell PC1, the cell generates a voltage in proportion to the amount of light. This proportional voltage is indicated by meter M1. Potentiometer R1 sets the meter sensitivity. The sensitivity of meter M1 can be anything up to 1 mA dc. However, if you prefer low light levels and long exposures, you should install a meter having a sensitivity of 500 μA or less.

To calibrate the enlarging meter, first make a good print from a No. 2 or No. 3 negative, and take care not to disturb the enlarger controls. Then integrate the light by placing a diffusing disc (ground glass) under the lens. Place PC1 on the easel and adjust potentiometer R1 for a convenient meter indication. The meter is now calibrated for the light exposure time used on the test print. For other negatives focus on the easel, place the diffuser under the lens, and adjust the lens diaphragm to the reference meter indication. The exposure time is the same as used for the test print.

PARTS LIST

Item	Description
M1	Meter, dc, 100-, 250-, or 500-μA
PC1	Solar cell, Allied Radio Shack 27-1710 (or equiv)
R1	Potentiometer, linear taper, 5000-ohm

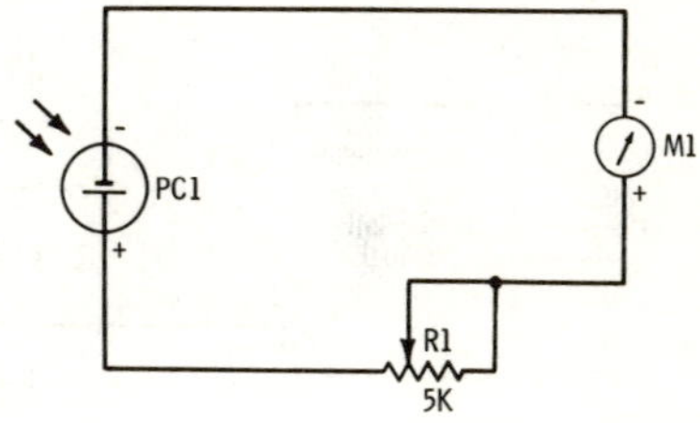

Fig. 55. This light meter can be used to establish the proper light and exposure time.

56 STOP-MOTION FLASH-TRIPPER

You, too, can take flash pictures the instant a pin pricks a balloon, a hammer breaks a lightbulb, or a bullet leaves the barrel of a gun.

You will need a miniamp—one of those amplifier modules rated 1 watt or less, and it must have an output transformer. Do not use an OTL amplifier.

The amplifier is terminated with a resistor at its highest output impedance, preferably 16 ohms. Potentiometer R1 is connected across this terminating resistor. Make certain that the connections to the sync terminals of the strobe light are correctly polarized. Positive terminal goes to the anode (A) of the SCR.

Darken the room, open the camera shutter, and strike a lightbulb with a hammer. The sound of the hammer striking the lightbulb will trigger the flash. Potentiometer R1 sets the overall sensitivity. Use the minimum sensitivity so that extraneous sounds do not trigger the system at the wrong time.

PARTS LIST

Item	Description
D1	Rectifier, silicon, 50-PIV, Motorola HEP-154 (or equiv)
Q1	Silicon controlled rectifier, GE C5B (or equiv)
R1	Potentiometer, 5000-ohm
R2	Resistor, 2700-ohm, ½-watt
Misc.	Microphone, ceramic

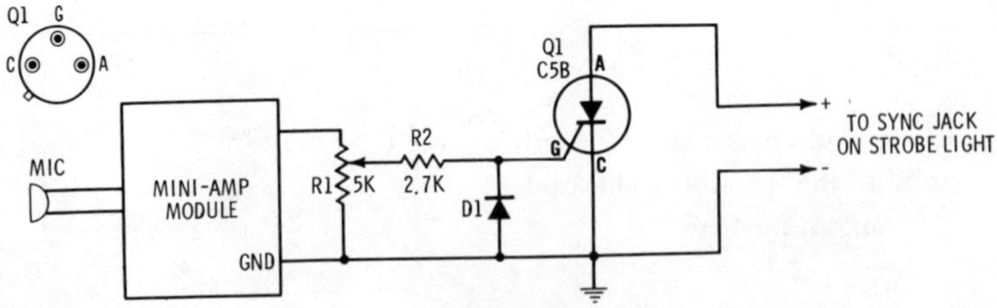

Fig. 56. This circuit will fire your strobe light at the right time to photograph fast motion.

POWER SUPPLIES

57 ZENER REGULATOR

When the output voltage from a power supply is too high for a solid-state project, cut it down to size with a zener-diode voltage regulator.

The value for R1 is calculated from the total current in the circuit (curent in the diode and current in the load) divided into the voltage drop across resistor R1. In Fig. 57, the values given in parentheses are calculated values used to develop a 9-volt at 50 mA output from a 12-volt source. These values are used as an example.

The value for R1 is 12 volts, minus the 9V, divided by the load current plus 10 percent of the load current.

$$R1 = \frac{12 - 9}{0.05 + 0.005} = \frac{3}{0.055} = 54.5 \text{ ohms}$$

Resistor R1 should be 55 ohms.

Wattage rating of the resistor is figured from the voltage and current ($P = E \times I = 3 \times 0.055 = 0.165$). Use at least ½ watt for resistor R1.

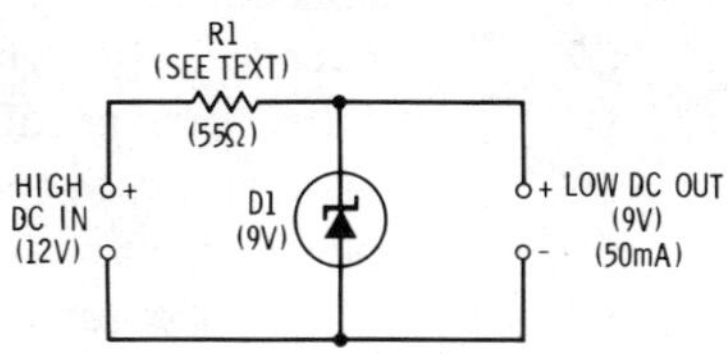

Fig. 57. The zener diode can be used to provide a regulated output voltage for solid-state projects or transistor radios.

The power rating for the zener diode must be figured on the basis of the total current because when the load is removed the diode will conduct the full 55 mA. Zener diode power is: $P = E \times I = 9 \times 0.055 = 0.495$ watt or ½ watt. For this application a 1-watt zener diode should be used. This is particularly important if the source voltage rises. For example, a car battery is 12 volts except when the motor is running, and then the voltage is about 14 volts. The zener current will increase to about 90 mA and this current will produce nearly 1 watt of heat.

58 REGULATED 9-VOLT POWER SUPPLY

Providing 9 volts at approximately 250 mA, this lab-type power supply will handle many experimenter projects. There are a number of 6.3-volt filament transformers that will provide approximately 9-volts *peak* when less than 500-mA output current is used. For best output regulation, use a 12-volt filament transformer.

You can change the zener diode to 6 or 12 volts (and possibly change the value of R1) and get a regulated 6- or 12-volt supply. For 12-volts output, you must use a filament transformer that is rated at 12 volts or higher.

Filtering is extremely good because the *electrical* filter capacitor equals the value of capacitor C2 times the gain of transistor Q1; this can add up the equivalent of thousands of microfarads.

PARTS LIST

Item	Description
C1	Capacitor, electrolytic, 500-μF, 25-Vdc
C2	Capacitor, electrolytic, 100-μF, 15-Vdc
D1	Bridge rectifier, 50-PIV, Motorola HEP-175 (or equiv)
D2	Zener diode, 9.1 volt, Motorola HEP-104 (or (equiv)
Q1	Transistor, npn, Motorola HEP-240 (or equiv)
R1	Resistor, 560-ohm, ½-watt
T1	Transformer, filament, 12-volt (see text)

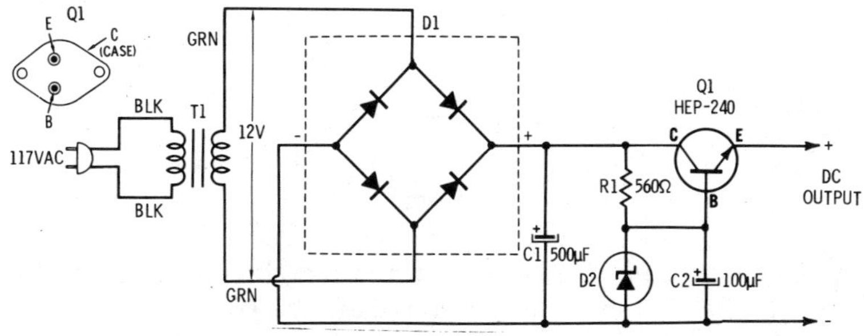

Fig. 58. Zener diode D2 can be changed to provide regulated output voltages from 6 to 12 volts.

59 BUDGET POWER SUPPLY FOR SOLID-STATE PROJECTS

There are many half-wave transformers available from the surplus market at rock-bottom prices. Use a bridge rectifier for full-wave rectification, and you have an easy-to-filter output.

The dc output voltage equals the secondary voltage (rms) rating of T1 times 1.4. Or working backwards, the secondary voltage must be 0.707 times the desired output voltage.

Silicon rectifiers D1 through D4 must have a PIV rating at least equal to the dc output voltage. The current rating of the diodes must be at least equal to the current requirements of the project being powered by the supply.

PARTS LIST

Item	Description
C1	Capacitor, electrolytic, 2500-μF (voltage rating higher than the output voltage)
D1, D2, D3, D4	Rectifier, silicon, 500-mA, 100-PIV
T1	Transformer, 117-Vac primary (secondary voltage equal to desired output times 0.707)

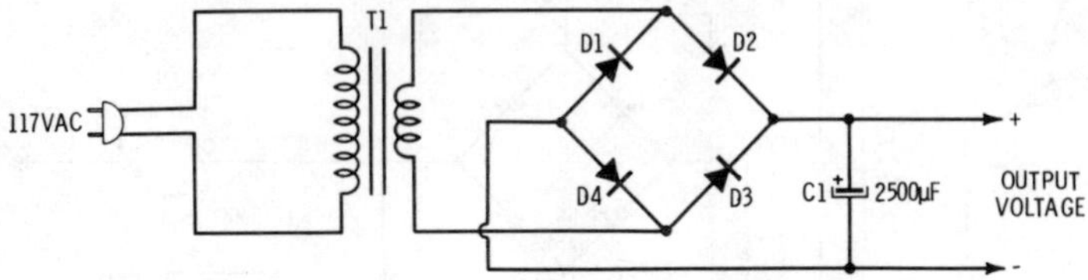

Fig. 59A. A bridge rectifier can be used to provide full-wave output from an untapped secondary winding.

Fig. 59B. The multitap primary and secondary windings of an Allied Radio series 54-4731 "low-voltage rectifier transformer" allows the experimenter to obtain any dc output between 7 and 40 volts in approximate 1.5- or 2-volt increments.

60 VOLTAGE DOUBLER

Found in many CB transceivers, this full-wave voltage doubler provides reasonably good regulation with a dc output voltage that is twice the applied ac voltage.

On the positive half-cycle, capacitor C1 is charged through diode D1. On the negative half-cycle, capacitor C2 is charged through diode D2. The dc output is the sum of the voltages across capacitors C1 and C2.

Capacitors C1 and C2 should be a minimum of 100 μF, and should be at least twice the peak ac voltage. The larger the capacity value, the greater will be the filtering and regulation. Silicon rectifiers D1 and D2 should have a PIV rating at least twice the dc output voltage.

PARTS LIST

Item	Description
C1, C2	Capacitor, electrolytic, 100-μF or larger (working voltage should be twice the maximum output voltage)
D1, D2	Silicon diode, 500-mA or larger (rated at a PIV at least twice the maximum dc output voltage)

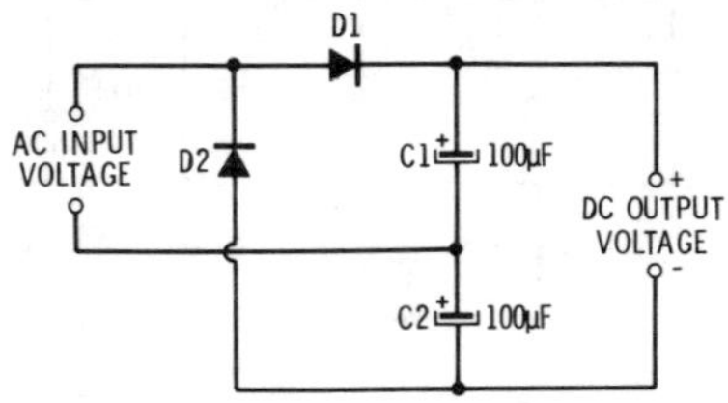

Fig. 60. A dc voltage value twice the value of the applied voltage is provided by this voltage-doubler circuit.

61 VOLTAGE TRIPLER

When you need a high voltage, but don't have the required power transformer, a voltage tripler might work. It provides a dc output approximately three times that of the applied ac voltage.

Capacitor C1 is approximately 8 to 20 microfarads at 150 volts; C2 and C3 should be a minimum of 100 microfarads at 250 volts. Larger values for capacitors C2 and C3, will improve the filtering.

On the negative half-cycle, capacitor C1 charges through diode D2, and capacitor C3 charges through diode D1. On the positive half-cycle, the charge on capacitor C1, plus the applied ac voltage, charges capacitor C2 through diode D3. The dc output is the sum of the voltages across capacitors C2 and C3, which is almost three times the applied voltage.

PARTS LIST

Item	Description
C1	Capacitor, electrolytic, 16-µF, 150-Vdc
C2, C3	Capacitor, electrolytic, 100-µF, 300-Vdc
D1, D2, D3	Rectifier, silicon, 1000-PIV Motorola HEP-160 (or equiv)

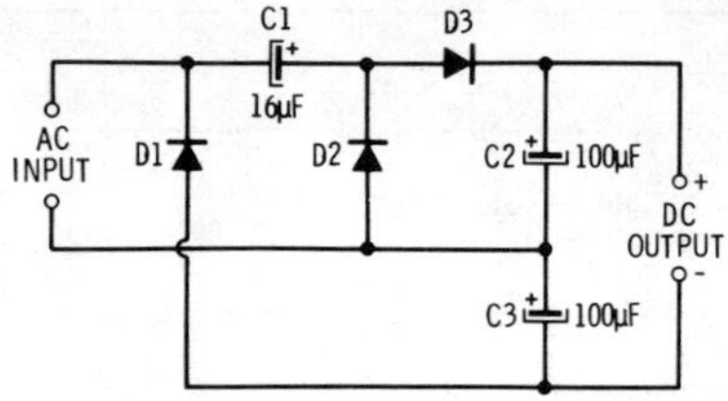

Fig. 61. This circuit will provide an output voltage that is three times the input voltage.

62 LOW-RIPPLE PREAMPLIFIER POWER SUPPLY

Just a few components are needed for a line-powered, low-voltage, low-current supply for audio preamplifiers. The values for different voltage and current outputs are given in the parts list.

For universal ease of application, diodes D1 and D2 are silicon rectifiers with a minimum of 200 PIV at any current rating.

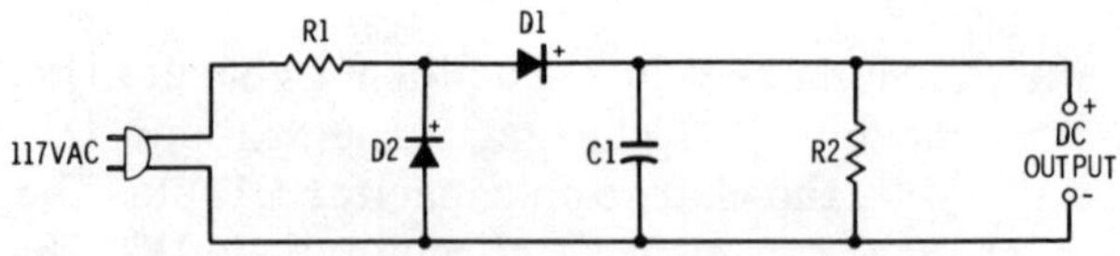

Fig. 62. Circuit of a power supply for low-current devices such as preamplifiers.

PARTS LIST

Item	Description
For 12-volt, 1-mA supply	
C1	Capacitor, electrolytic, 250-μF, 15-Volt
D1, D2	Rectifier, silicon, 200 PIV
R1	Resistor, 43,000-ohm, ½-watt
R2	Resistor, 180,000-ohm, ½-watt
For 12-volt, 2-mA supply	
C1	Capacitor, electrolytic, 250-μF, 15-Volt
D1, D2	Rectifier, silicon, 200 PIV
R1	Resistor, 22,000-ohm, ½-watt
R2	Resistor, 100,000-ohm, ½-watt
For 25-volt, 2-mA supply	
C1	Capacitor, electrolytic, 250-μF, 30-Volt
D1, D2	Rectifier, silicon, 200 PIV
R1	Resistor, 18,000-ohm, ½-watt
R2	Resistor, 180,000-ohm, ½-watt

RADIO RECEIVERS

63 DIRECT-COUPLED RADIO

A shirt-pocket project, the direct-coupled radio in Fig. 63A uses transistor Q1 as a diode detector *and* first audio amplifier. Detection occurs across the base-emitter junction, which functions as a diode. Normal base-emitter capacitance provides the rf filtering. The detected modulation is amplified by transistor Q1 and further amplification in accomplished by transistors Q2 and Q3.

PARTS LIST

Item	Description
B1, B2	Battery, 1.5-volt "AA" or "AAA" cells
C1	Capacitor, variable, 365-pF
L1	Antenna, tapped coil on ferrite rod
Q1, Q3	Transistor, npn, Motorola HEP-641 (or equiv)
Q2	Transistor, pnp, Motorola HEP-253 (or equiv)
R1	Potentiometer, 5000-ohm
R2	Resistor, 100-ohm, ½-watt
Misc.	Headphones, magnetic

Coil L1 can be tapped (transistor type) ferrite antenna coil. Tuning capacitor C1, shown at the right in Fig. 63B, is a miniature *poly*-type variable. The earphone can be a magnetic or crystal type as long as the impedance is in the 2500- to 5000-ohm range. Potentiometer R1 is adjusted for the best sound output—least distortion consistent with maximum volume.

During construction of this project take particular care not to get the pnp and npn transistors in the wrong positions.

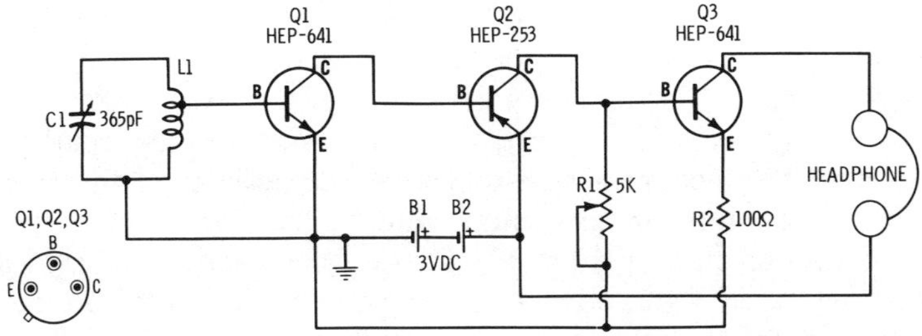

Fig. 63A. A pocket-size, three-transistor radio.

Fig. 63B. In order to make the direct-coupled radio pocket size, a *poly* type variable capacitor—shown held between the fingers—must be used. A standard size single-gang capacitor as shown on the left can be used when overall size is unimportant.

64 AMPLIFIED CRYSTAL RADIO

That old favorite, the crystal radio, becomes more than just a weak voice buried in the headphone when the audio is amplified with a "junk box" amplifier.

Transistor Q1 can be just about any general purpose pnp germanium type such as the 2N107, 2N109, etc. The RCA SK3003 specified in the parts list gives a little extra gain.

L1 is any adjustable antenna coil designed for broadcast band use. The headphones must be a magnetic type to act as collector load and for maximum volume. To align the receiver, set the dial of tuning capacitor C1 to the frequency of a local broadcast station and adjust the slug in coil L1 until you hear the station in the phones. For reception of weaker signals, the receiver should be connected to an *earth ground* such as a cold water pipe.

To feed the radio output into an amplifier and speaker, replace the earphone with a 1000-ohm, ½-watt resistor and connect a 0.1-μF, 25-Vdc capacitor from the collector of the transistor to the amplifier input. Then connect together the radio and amplifier grounds.

PARTS LIST

Item	Description
B1	Battery, 9-volt, Type 2U6 (or equiv)
C1	Capacitor, variable, 365-pF
C2	Capacitor, 0.2-μF, 10-Vdc
D1	Diode, 1N60
L1	Antenna coil adjustable for broadcast band
Q1	Transistor, pnp, RCA SK3003 (or equiv)
R1	Resistor, 100,000-ohm, ½-watt
Misc.	Headphones, magnetic

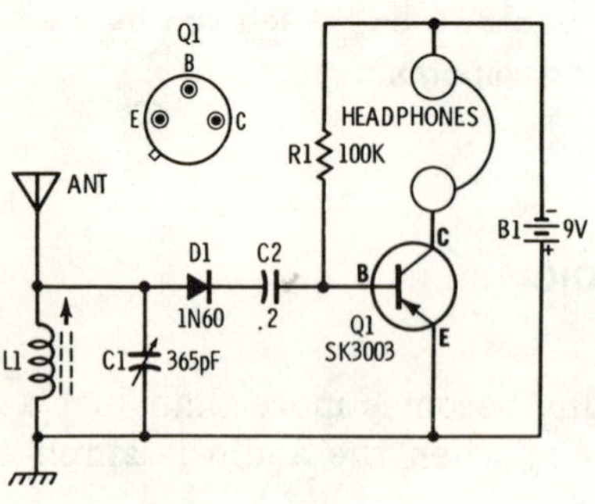

Fig. 64. A transistor is used to boost the volume of the signal from the detector.

65 BFO FOR TRANSISTOR RADIOS

Placed near a multiband transistor radio, this *radiating* bfo allows reception of cw and ssb signals in addition to the normal reception.

The bfo is a Hartley oscillator that is tunable across the broadcast band. Oscillator harmonics extend to the higher short-wave frequencies where they "beat" against cw and ssb signals, producing a tone signal in the speaker.

Once capacitor C2 is adjusted for the proper beat frequency, the bfo is positioned near the transistor radio for optimum reception. No antenna is needed if the unit is assembled in a plastic cabinet.

Note that the tap on antenna coil L1 is closest to the bottom end. The oscillator will not work if coil L1 is reversed so that the tap is near the top.

PARTS LIST

Item	Description
C1, C3, C4	Capacitor, 0.05-μF, 25-Vdc
C2	Capacitor, variable, 365-pF
L1	Antenna coil, tapped (broadcast band)
Q1	Transistor, npn, GE-5 (or equiv)
R1	Resistor, 2200-ohm, ½-watt
R2	Resistor, 68-ohm, ½-watt

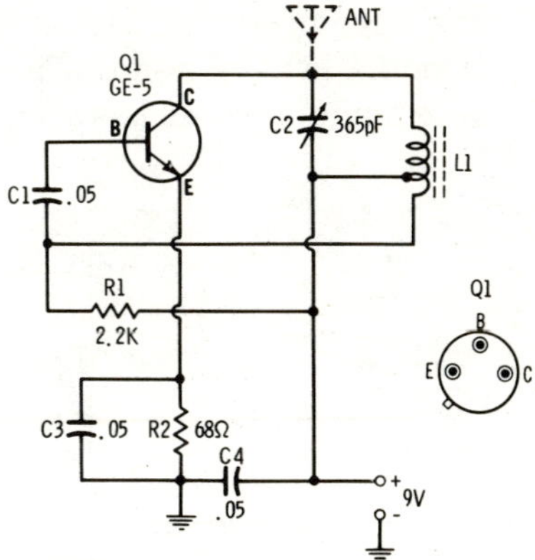

Fig. 65. Listen to cw and ssb signals by using this external BFO.

REMOTE CONTROLS

66 LIGHT TO CURRENT CONTROL

Heavy direct current is easily controlled without the use of massive power switches and wiring by using a LASCR (light-activated silicon controlled rectifier) between the control and the controlled circuits. The LASCR is similar to an SCR, except the LASCR gate is triggered by light rather than by current or voltage.

PARTS LIST

Item	Description
I1	Bulb or pilot lamp (see text)
Q1	Light-activated silicon controlled rectifier, GE (see text)
R1	Resistor, 47,000-ohm

The trigger source, I1, can be any flashlight lamp powered by two "D" cells. When the lamp is turned *on,* the LASCR gate is closed, and current is switched to the load.

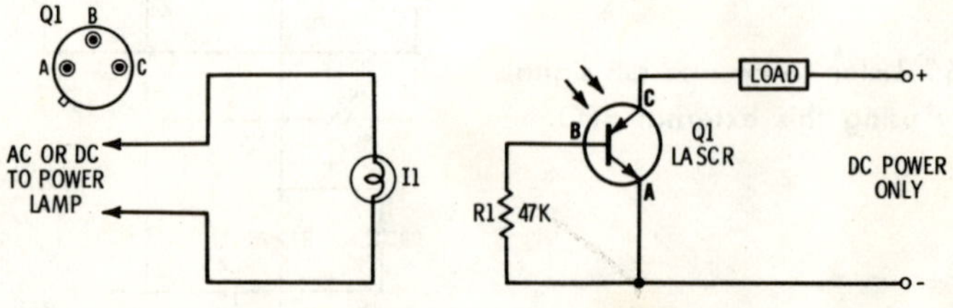

Fig. 66. The LASCR can control a heavy current using light as the trigger.

A suitable LASCR is one from the General Electric L8B series. Inexpensive LASCR units are occasionally available from surplus dealers. Be sure that the LASCR unit has the proper PIV rating for your application.

67 LOW VOLTAGE RELAY CONTROL

Using ordinary bell wire, you can safely control a remote 120-Vac power source. The secret behind it all is a unique hysteresis relay (K1). Normally, the coil of relay K1 represents a high impedance, so there is practically no current in the coil, there is no magnetic field, and contacts of relay K1 are open. When switch S1 closes the loop on the hysteresis coil, the impedance of the relay coil drops, there is current and a magnetic field, and the armature of K1 pulls in—closing the contacts.

When switch S1 is open, the voltage across the hysteresis coil is approximately 30 volts. When S1 is closed, the current through the hysteresis coil is extremely small. The relay is safe enough that ordinary bell or hook-up wire can be used for control leads.

PARTS LIST

Item	Description
K1	Relay, hysteresis (Alco)
S1	Switch, spst
Misc.	Bell wire

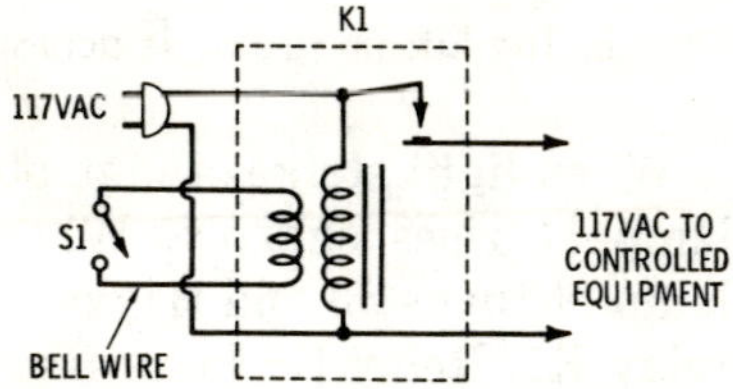

Fig. 67. This special relay permits simple control of ac power with ordinary bell wire.

68 LIGHT-CONTROLLED LATCHING SWITCH

A flashlight beam stabs out and the sound for an irritating television commercial vanishes. Moments later, when the program returns, the flashlight beam stabs out again, and the sound snaps back *on*. Between the amplifier and the speaker is the light-controlled latching switch.

PARTS LIST

Item	Description
D1	Rectifier, silicon, 200-PIV
I1	Lamp, neon, NE-83
K1	Relay, latching, Guardian IR-610L-A115 (or equiv)
PC1	Photocell, Clairex CL505 (or equiv) for high light level; CL704 or CL705 (or equiv) for low light level
R1	Resistor, 22,000-ohm, ½-watt
R2	Potentiometer, 1-megohm
R3	Resistor, 100-ohm, ½-watt
Q1	Silicon controlled rectifier, GE C6B (or equiv)

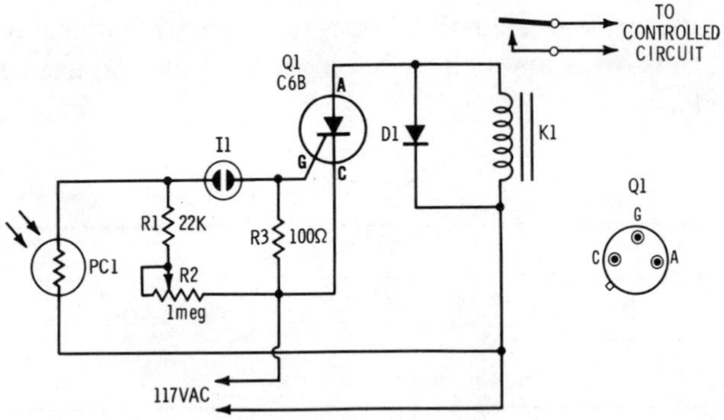

Fig. 68. The latching relay is actuated by shining a light on the photo cell.

When light strikes the photocell PL1 the voltage across neon lamp I1 rises sharply. When conduction voltage is reached, lamp I1 turns *on* and triggers the SCR, which in turn activates relay K1. Relay K1 is an impulse relay whose contacts stay in position after coil current is removed. So the first impulse opens the contacts of K1 and the second impulse closes them, etc.

To prevent ambient light from tripping the photocell, PC1 should be recessed at least an inch inside a cardboard or metal tube.

TEST EQUIPMENT (GENERAL)

69 APPLIANCE TESTER

A simple circuit consisting of a 50-watt lightbulb, a 15 ampere fuse, and a power outlet is all that's needed to check out appliances such as toasters and electric coffeepots.

To check for open circuits, first plug the tester into a live outlet. Then connect the test leads to the power cord of the appliance; if the bulb lights, the circuit is good (not open). Because the appliance is in series with the lightbulb, the bulb won't light to full brilliance. You are interested only in whether the bulb lights, rather than in the level of brilliance.

If you suspect that there is a short between the appliance motor or heating coil and the frame—which is a shock hazard —connect one test lead to the frame and connect the other test lead in turn to each of the terminals on the power cord of the appliance. If the bulb lights with either connection, there is a short to the frame.

PARTS LIST

Item	Description
F1	Fuse to match load
I1	Lamp, 50-watt
PL1	Receptacle, ac

After the repair is made, try out the appliance using the fused power outlet (PL1). This way, if the appliance is still defective, it will blow fuse F1 rather than a fuse or breaker in the basement.

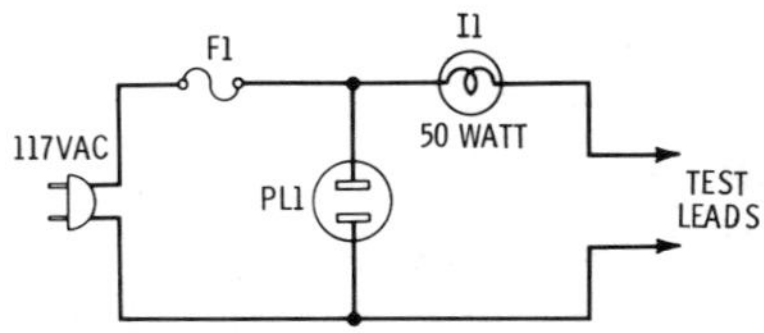

Fig. 69. A safe way to test appliances is to use this circuit.

70 RF PROBE FOR VOMs

Assemble this accessory in a metal can, add a shielded cable, and you can make relative measurements of rf voltages up to 200 MHz on a 20,000-ohms-per-volt multimeter. The rf input voltage must not exceed 30 volts, because this is the breakdown voltage of the 1N60 diode.

PARTS LIST

Item	Description
C1	Capacitor, 500-pF, 400-Vdc
C2	Capacitor, electrolytic, 0.001-μF, 100-Vdc
D1	Diode 1N60
R1	Resistor, 15,000-ohm, ½-watt

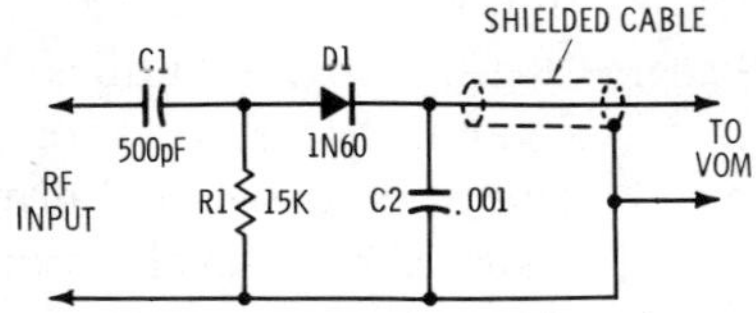

Fig. 70. Measure rf voltages with a VOM by using this circuit.

71 RF PROBE FOR VTVMs

Only three components are needed to make a VTVM measure rf voltage up to 200 MHz (depending on the diode used). The probe should be built in a metal can with shielded wire for the connecting lead to the VTVM. The shield must connect to the metal can.

The output voltage from the probe is positive and the VTVM will indicate the peak value of the rf voltage. To determine the rms value, multiply the VTVM reading by 0.707.

The rf input voltage must not be over 30 volts because this is the breakdown voltage of the 1N60 diode.

PARTS LIST

Item	Description
C1	Capacitor, disc, 50-pF, 500-Vdc
D1	Diode, 1N60
R1	Resistor, 20-megohm, ½-watt

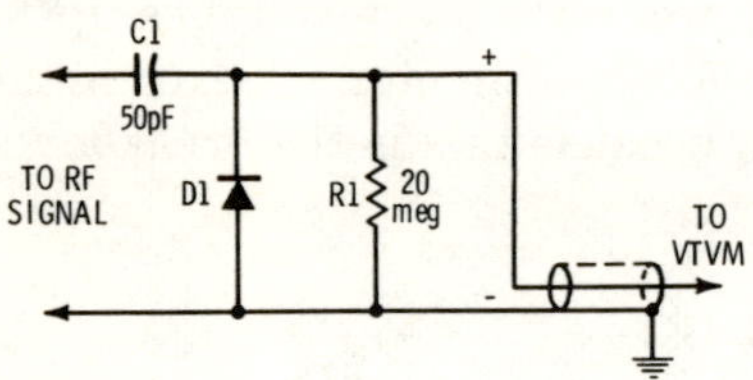

Fig. 71. With this circuit, your VTVM can measure frequencies up to 200 MHz.

72 OSCILLOSCOPE TIME-BASE CALIBRATOR

Operating on exactly 100 kHz, the scope calibrator provides a precise reference for calibrating the variable time base oscillator of general purpose scopes. If the scope is set, for example, so one cycle of the calibrator output fills exactly 10 graticule divisions, then each division will represent 1 MHz or 1 microsecond. If the scope is adjusted to display 10 cycles on

10 graticule divisions, then each division will represent 100 kHz or 10 microseconds. If the scope time base is sufficiently stable, you can make precise measurements of pulse width, length, and frequency.

PARTS LIST

Item	Description
C1, C3	Capacitor, 0.01-μF, 25-Vdc
C2	Capacitor, 0.002-μF, 25-Vdc
Q1	Transistor, npn, Motorola HEP-641 (or equiv)
R1	Resistor, 100,000 ohm, ½-watt
R2	Resistor, 1000-ohm, ½-watt
X1	Crystal, 100-kHz

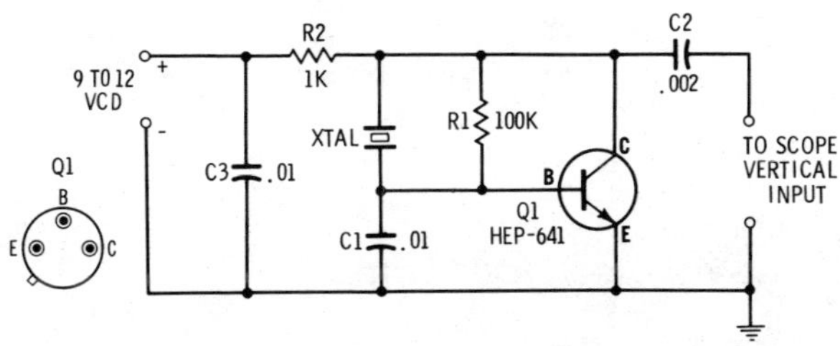

Fig. 72. This circuit can be used to set an accurate time base on your scope.

73 VERTICAL-INPUT CALIBRATOR

Back-to-back zener diodes provide the vertical input calibrator of a scope with a zero-dc reference output. Whether the calibration voltage is fed to the ac or dc input of a scope, the baseline will remain fixed.

When the anode of diode D1 goes positive, diode D1 conducts in the forward direction. The reverse voltage across diode D2 rises to 5 volts, and then is clamped at 5 volts by the zener action of diode D2. The reverse action takes place when the anode of diode D1 goes negative. The result is a 10-volt peak-to-peak square wave that can be used to calibrate the vertical input of the scope.

PARTS LIST

Item	Description
D1, D2	Zener diode, 5-volt, 250-mW
R1	Resistor, 270-ohm, ½-watt

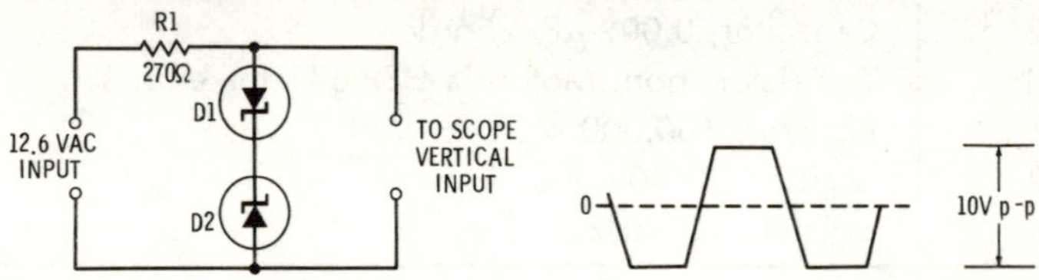

Fig. 73. This simple circuit can be used to calibrate the vertical deflection on an oscilloscope.

74 LOW-VOLTAGE DIODE TESTER

Low-voltage signal-type diodes are easily checked with a go/no-go tester. The only restriction on its use is that the diode being checked must be rated to handle 60 mA. Diodes such as the 1N34 cannot be checked because the test current is too high.

If the diode is good, the lamp will light with the diode connected in one direction, and remain dark when the diode is reversed. If the lamp stays on when the diode is reversed, the diode is shorted. If the lamp remains dark when the diode is connected in both directions, the diode is open.

PARTS LIST

Item	Description
B1	Battery, 6V (four "C" or "D" cells)
I1	Lamp, No. 49
R1	Resistor, 68-ohm, ½-watt

Fig. 74. Test low-current diodes with this low-voltage tester.

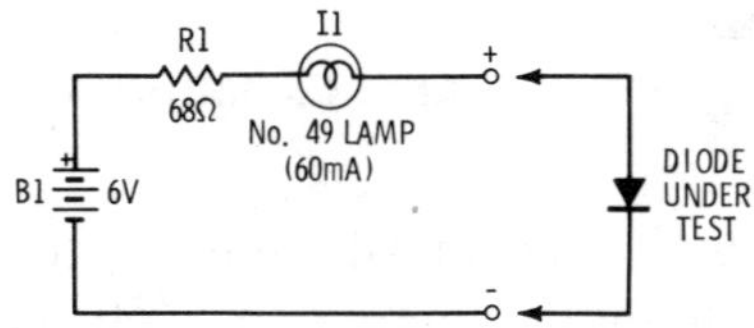

75 SILICON-RECTIFIER CHECKER

Using the go/no-go principle, this silicon-rectifier checker spots defective diodes before they are connected into the circuit. This checker is intended only for silicon rectifiers rated higher than 200 mA, and indicates short or open conditions.

The lamp must be as specified: 120 volts at 15 watts. If you use a larger lamp, the rectifier might be destroyed.

Close switch S1 to check the lamp by turning it on. Then connect the rectifier diode under test both ways, opening S1 for each test. One direction should cause the lamp to remain lighted; reversing the diode should cause the lamp to go out. If the lamp stays on in both directions, the diode is shorted. If the lamp is dark in both directions, the diode is open.

PARTS LIST

Item	Description
D1	Diode, 750 mA, 200 PIV
I1	Lamp, 125 volt, 15 watt
S1	Switch, spst (or push button)
Misc.	Line cord and ac plug

Fig. 75. Circuit for testing rectifiers with a voltage rating over 200 volts.

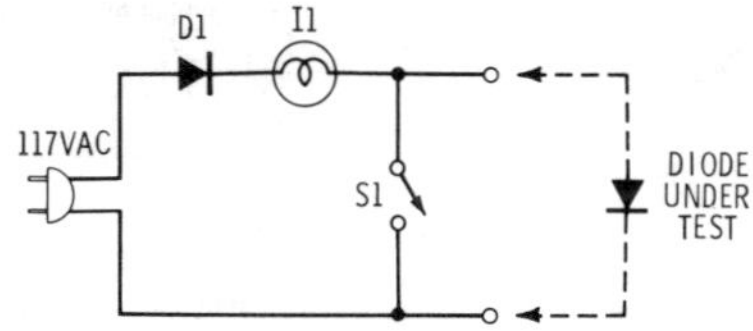

76 NICKEL-CADMIUM BATTERY CHARGER

Providing an adjustable output voltage up to 35 Vdc and a maximum current of 750 mA, this battery charger handles just about any nickel-cadmium battery used in consumer and experimenter equipment.

Transistor Q1 must be mounted on a heat sink, which can be the cabinet. But since the collector of Q1 is connected to the transistor case, the transistor must be insulated from the metal cabinet with the insulating hardware provided in a transistor mounting kit. For best heat dissipation, spread a layer of silicone transistor-mounting grease on both sides of the mica insulator.

When charging one cell or a string of series-connected nickel-cadmium cells, connect an ammeter in series with the charger and adjust to the current specified for the battery(s). Never attempt to rapid-charge nickel-cadmium batteries (unless so

PARTS LIST

Item	Description
C1	Capacitor, electrolytic, 100-μF, 50-Vdc
D1	Rectifier, silicon, 750-mA, 100-PIV
Q1	Transistor, power, 40-watt, pnp (any type)
R1	Potentiometer, 2000-ohm
T1	Transformer, filament, 24-Vac secondary, 117-Vac primary

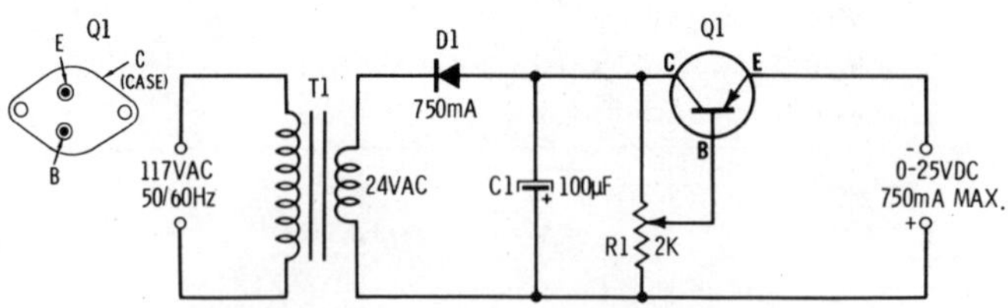

Fig. 76. Circuit for an adjustable battery charger for nickel-cadmium batteries.

designed), because excess charging current can result in permanent damage.

Note: The output current must be limited to the current rating of diode D1. In this case, about 750 mA. Short circuit protection can be provided by placing a fuse in the lead to the batteries being charged.

77 LIGHTBULB CHARGER FOR FLASHLIGHT BATTERIES

This circuit in a fancy commercial package will cost around $5, but 50 cents should cover all costs if you build it yourself.

The lamp maintains a constant charge current of approximately 20 mA on all batteries, or series-connected batteries, up to 22.5 volts.

Give small flashlight batteries ("AA" or "AAA") about 10 hours charge; give the "C" and "D" cells about 20 hours charge. You can also recharge nickel-cadmium cells stamped with a charge rate of approximately 20 to 25 mA.

PARTS LIST

Item	Description
I1	Lamp, candelabra, No. S6, 6-watt
D1	Rectifier, silicon, 200-PIV, 100-mA minimum

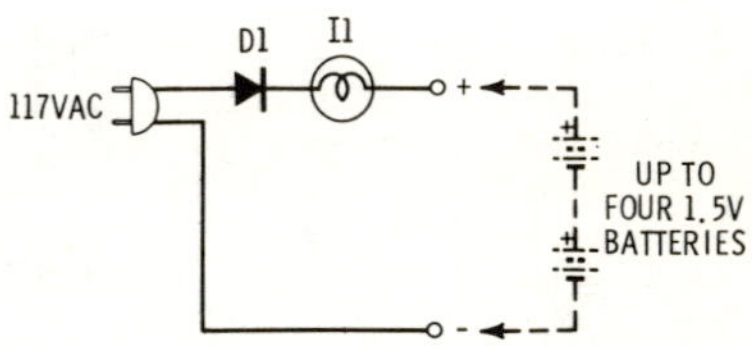

Fig. 77. This charger uses a light bulb to limit the charging current.

TEST EQUIPMENT (AUDIO)

78 MICROPHONE BEEPER

You can always feed a signal generator into a microphone input to check out an audio system, but how do you check the microphone itself Saying "woof, woof, hello test" gets mighty tiring. Instead, clamp the microphone beeper to the front of the microphone with a rubber band, and you'll send a continuous

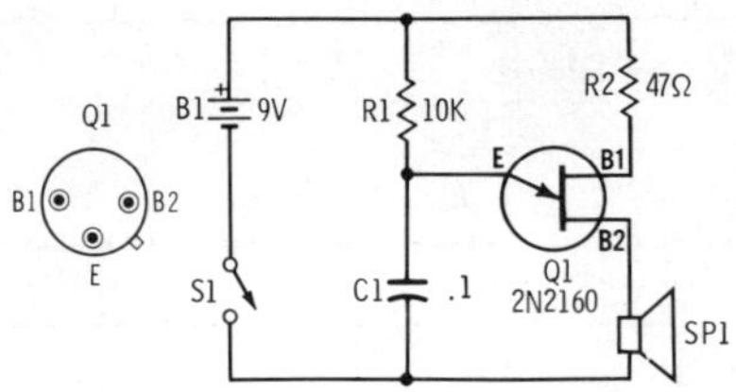

Fig. 78. Here is an easy way to check out your microphone.

PARTS LIST

Item	Description
B1	Battery, 9-volt, Type 2U6 (or equiv)
C1	Capacitor, 0.1-μF, 10-Vdc
Q1	Transistor, unijunction, 2N2160
R1	Resistor, 10,000-ohm, ½-watt
R2	Resistor, 47-ohm, ½-watt
SP1	Speaker, miniature, 3.2- or 8-ohm

tone signal *through the microphone.* The continuous tone signal lets you take your time checking the microphones, connecting cables, etc.

The beeper can be built in a small plastic case—nothing is critical. The speaker can be any size from one to three inches.

79 AMPLIFIED VU METER

You can ride gain on audio signals, just as the professionals do, with an amplified volume-unit (VU) meter.

The circuit shown will produce meter indications on signal voltages almost down to microphone level. Sensitivity control R1 allows the unit to handle even high power levels. Emitter-follower Q1 presents a high impedance to the input terminals (across R1), and emitter-follower Q3 provides a low impedance output for VU meter M1. Meter M1 can be any VU meter such as an inexpensive miniature type or a professional model.

Power for the project can be supplied by two series-connected, Type 2U6 9-volt batteries.

The amplified VU meter can also be used as a signal tracer by replacing meter M1 with headphones that have an impedance in the 500- to 3000-ohm range.

PARTS LIST

Item	Description
C1, C3	Capacitor, 0.1-μF, 25-Vdc
C2	Capacitor, electrolytic, 10-μF, 15-Vdc
C4	Capacitor, electrolytic, 30-μF, 15-Vdc
M1	Meter, VU
Q1, Q3	Transistor, pnp, Motorola HEP-250 (or equiv)
Q2	Transistor, pnp, Motorola HEP-252 (or equiv)
R1	Potentiometer, audio taper, 1-megohm
R2, R8	Resistor, 470,000-ohm, ½-watt
R3, R9	Resistor, 4700-ohm, ½-watt
R4	Resistor, 120,000-ohm, ½-watt
R5, R6	Resistor, 10,000-ohm, ½-watt
R7	Resistor, 1500-ohm, ½-watt

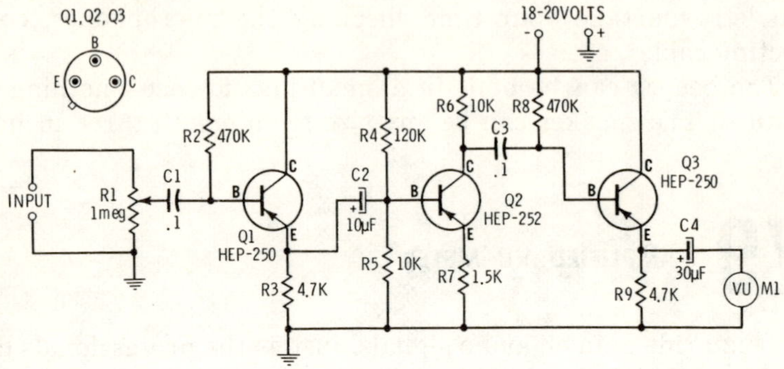

Fig. 79. Keep a close check on the audio signal with an amplified VU meter.

80 STEREO BALANCER

It looks ridiculously simple, but this instrument will give you precise volume and tone control balance between the left and the right stereo amplifiers.

For maximum convenience, meter M1 should be a zero-center type. The resistors should be matched to within 5 percent and the diodes should be a matched pair. Note that the leads from the diodes connect to the "hot" or ungrounded terminal on the amplifier.

PARTS LIST

Item	Description
D1, D2	Diode, 1N60
M1	Meter, zero center, 500-0-500 μA dc
R1, R3	Resistor, 560-ohm, ½-watt
R2	Resistor, 1000-ohm, ½-watt

Optimum stereo level and phase balance occur for matched speakers when the meter indicates "0." If meter M1 indicates either side of "0," the volume levels are not matched, or the

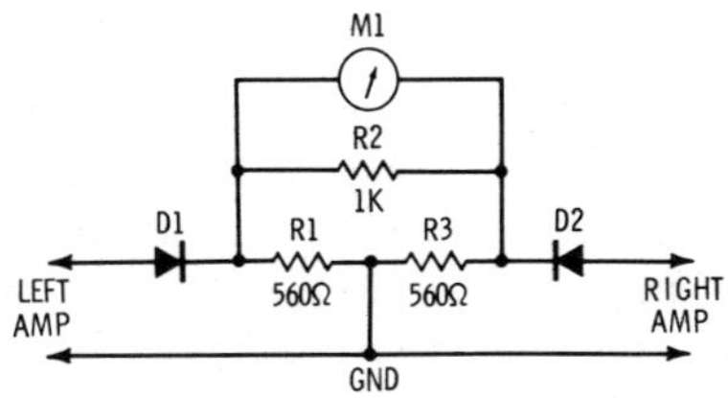

Fig. 80. This circuit and a zero-center meter can be used to accurately balance any stereo system.

wire are incorrectly phased. Check the phasing by double-checking the meter connection to the amplifier terminals.

An ordinary 0-1 mA dc meter can be substituted for M1, but keep in mind that the pointer can be driven off-scale in the reverse direction.

81 DISTORTION METER

This 1-kHz distortion meter is extremely accurate for measuring the distortion of power amplifiers.

Resistor R_x is the load resistor for the amplifier and should be 4, 8, or 16 ohms at the appropriate power rating. The ac meter used with this circuit can be an ac-VTVM or a 5000 ohm-per-volt VOM (set to ac range). A sine wave generator set to produce a 1-kHz signal is connected to the input of the amplifier.

Adjust the amplifier volume to the desired power output. Set switch S1 to the calibrate position, and note the meter indication. Then set S1 to the total harmonic disortion (THD) position, and adjust coil L1 and resistor R1 for *minimum* indication on the meter.

The percent THD is equal to the *minimum* indication divided by the *calibrate* indication times 100.

The circuit works by filtering out the 1-kHz fundamental signal with the T-notch filter. The total harmonic content is not filtered out, and this is the signal indicated on the meter.

PARTS LIST

Item	Description
C1, C2	Capacitor, 0.01-μF, 100-Vdc, 5%
L1	Inductor, variable, UTC VC-15 (or equiv)
R1	Potentiometer, 25,000-ohm
R_x	Resistor, amplifier load (see text)
S1	Switch, dpdt

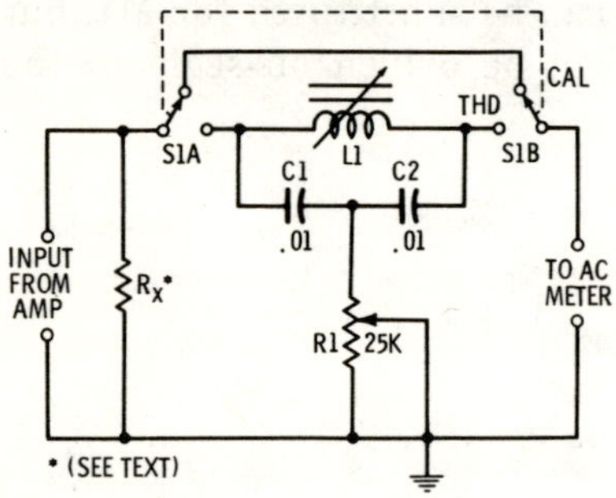

Fig. 81. Basic circuit required to measure harmonic distortion in an amplifier.

82 SINE-WAVE SQUARER

Two diodes connected in reverse-parallel can be used as an emergency square-wave generator. Since a germanium diode has a forward-conduction voltage of approximately 0.2 volt, any sine wave applied to the diodes will be clipped at 0.2 volt. The positive and negative half-cycles of the signal input add, giving a squared output of 0.4 volt peak-to-peak. The result is not a pure square wave because the rise and fall of the wave form follows the normal slope of the sine wave.

PARTS LIST

Item	Description
D1, D2	Diode, 1N60
R1	Resistor, 1000-ohm, ½-watt

To prevent excessive loading and possible distortion of the sine-wave input, a 1000-ohm resistor should be connected between the signal generator and the diodes as shown in Fig. 82.

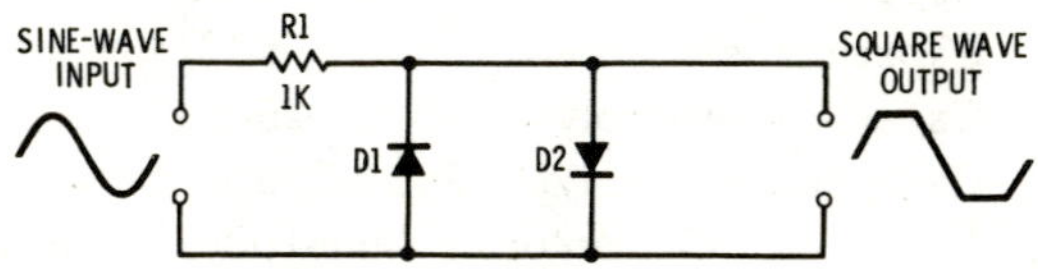

Fig. 82. Two diodes can be used to make an emergency square-wave generator.

83 AUDIO SNIFFER

If you have servicing problems in audio equipment, sniff them out quickly with an audio signal tracer. The sniffer has enough gain to overdrive the headphones with the output from a microphone or magnetic pickup. Transistor Q1 is a field-effect transistor (FET) which allows the use of a very high impedance volume control (R1). Control R1 can be any value up to 10 megohms, thereby effectively nullifying any loading effects on crystal microphones and pickups.

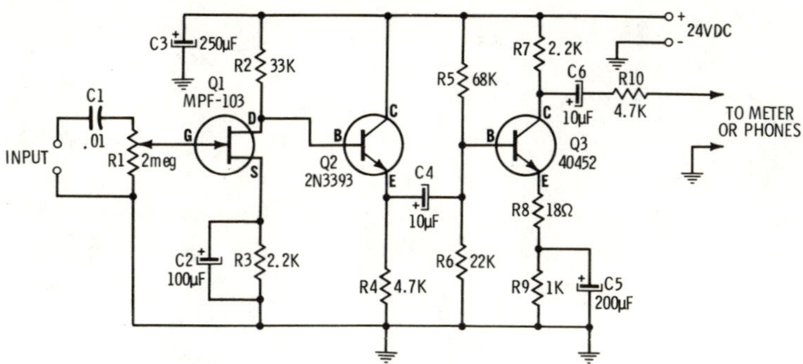

Fig. 83. Circuit for a highly sensitive audio-signal tracer.

PARTS LIST

Item	Description
C1	Capacitor, 0.01-μF, 400-Vdc
C2	Capacitor, electrolytic, 100-μF, 6-Vdc
C3	Capacitor, electrolytic, 250-μF, 50-Vdc
C4, C6	Capacitor, electrolytic, 10-μF, 25-Vdc
C5	Capacitor, electrolytic, 200-μF, 6-Vdc
Q1	Transistor FET, Motorola MPF-103 (or equiv)
Q2	Transistor npn, 2N3393
Q3	Transistor npn, RCA 40452 (or equiv)
R1	Potentiometer, audio taper, 2-megohm
R2	Resistor, 33,000-ohm, ½-watt
R3, R7	Resistor, 2200-ohm, ½-watt
R4, R10	Resistor, 4700-ohm, ½-watt
R5	Resistor, 68,000-ohm, ½-watt
R6	Resistor, 22,000-ohm, ½-watt
R8	Resistor, 18-ohm, ½-watt
R9	Resistor, 1000-ohm, ½-watt

TEST EQUIPMENT (RF)

84 CB SCOPE BOOSTER

Critical inspection of a transmitter signal and accurate measurement of modulation are possible only with an oscilloscope. Unfortunately, the rf output of a CB transmitter is so low that the scope pattern is barely discernible—unless you use a booster.

Since the vertical-deflection plate connections of a scope are at a high impedance, the plates can be connected to a parallel resonant circuit without causing much of a loading effect on the circuit. Using the resonant circuit shown, L1 and L2, a CB transmitter will just about fill a 5-inch oscilloscope screen, with a virtually insignificant amount of power loss.

First, wind coil L2 on the center of a ⅜-inch slug-tuned coil form. Then wind L1 adjacent to the ground end of coil L2. Connect L1 directly across the transmission line of the transmitter (with the antenna connected). Adjust the coil slug for minimum standing-wave ratio (SWR). If the coil is correctly made, there will be no change in the SWR of the antenna system. Adjust capacitor C3 for the desired height of the oscilloscope display. It might be necessary to readjust the slug each time capacitor C3 is adjusted. Note that you must use the *vertical-deflection plate* connections on the scope. The rf signal cannot pass through the vertical amplifiers of normal service-type oscilloscopes.

PARTS LIST

Item	Description
C1, C2	Capacitor, silver mica, 5-pF
C3	Capacitor, trimmer, 45-pF
L1	3 turns No. 22 solid, plastic insulated wire, adjacent to ground end of L2
L2	4 turns No. 18 enameled wire centered on a ⅜-inch rf slug-tuned coil form

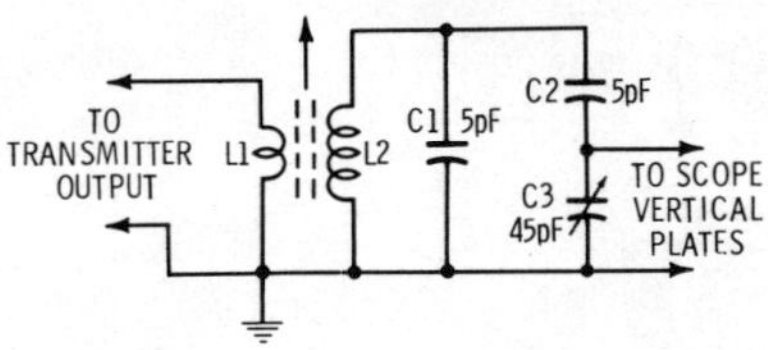

Fig. 84. The output signal from a CB transmitter can be viewed on an oscilloscope by using this circuit.

85 I-F SIGNAL GENERATOR

Using a 455-kHz crystal, this generator provides a signal source for testing and aligning a-m radio i-f circuits. The unit is built on perforated board or some other rigid mounting to achieve good circuit stability. A metal cabinet should be used to reduce radiation so that the signal level fed to the receiver will be determined primarily by control R2.

PARTS LIST

Item	Description
C1	Capacitor, 0.05-μF, 25-Vdc
C2	Capacitor, silver mica, 50-pF
C3	Capacitor, silver mica, 15-pF
L1	Coil, rf, 3.4- to 5.8-MHz Miller 21A473RB1 (or equiv)
Q1	Transistor, npn, GE-5 (or equiv)
R1	Resistor, 330,000-ohm, ½-watt
R2	Potentiometer, 5000-ohm
X1	455-kHz crystal

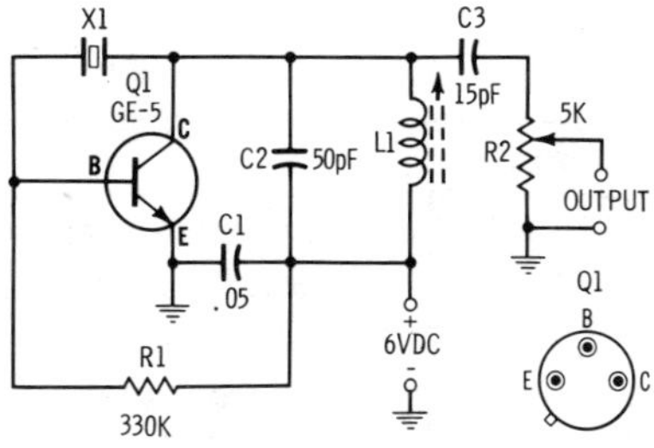

Fig. 85. Signal generator for aligning 455-kHz i-f stages.

To align the receiver, adjust the slug in coil L1 for maximum S-meter indication while feeding the output of the generator into a receiver, or connect the output to an oscilloscope and adjust coil L1 for maximum signal.

Turn the 6-volt power source on and off several times to make certain the oscillator starts consistently. If the oscillator fails to start every time, adjust L1 slightly until you obtain immediate and consistent starting each time the power is applied.

86 FM ALIGNMENT OSCILLATOR

We do not suggest that you start aligning your stereo fm receiver, but some budget-priced and early-model monophonic receivers can be aligned or peaked with this fm alignment oscillator. The oscillator provides a radiated signal within 10 feet of the receiver. It is strong enough for alignment purposes, yet it will not overload the front end.

Coil L1 must be made with extra care. The form for L1 is a ⅜-inch dowel or drill and the 4-turn section is tight-wound with No. 18 wire and no spacing between turns. The 3-turn section is stretched after winding—to a length of ⅜ in. from the tap to the end of the coil. The tap is made by scraping off some enamel from the No. 18 enameled wire, tinning the bare area, and then soldering a solid wire to the tinned area.

The frequency of the oscillator is set by adjusting trimmer capacitor C4.

PARTS LIST

Item	Description
B1	Battery, mercury, 7 volt, TR175 (or equiv)
C1, C2	Capacitor, 500-pF, 100-Vdc
C3	Capacitor, silver mica, 5-pF
C4	Capacitor, trimmer, 2.7- to 30-pF
L1	Made of No. 18 enameled wire (see text)
Q1	Transistor, pnp, Motorola HEP-637 (or equiv)
R1	Resistor, 100,000-ohm, ½-watt
R2	Resistor, 470-ohm, ½-watt

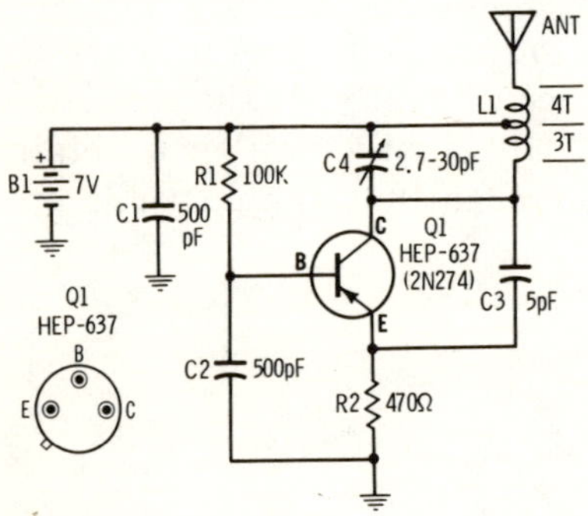

Fig. 86. Align that old fm receiver with this low cost signal generator.

87 SIGNAL INJECTOR

This multivibrator produces pulses at approximately 1 kHz; the harmonics of these pulses extend as high as 14 MHz. This multivibrator is useful for servicing both audio and rf circuits.

The device is a signal injector, meaning that you inject the signal into a radio or amplifier by starting at the speaker and working back to the input or antenna. The difficulty is localized when you can no longer hear the injected signal in the speaker.

Precaution: The signal level is quite high and can destroy a small signal transistor. Therefore, when working on a small-signal stage where the signal is to be injected into a transistor base, make an inductive connection by insulating the test probe with a layer of tape. Rest the probe against the base lead. This will provide adequate signal for sensitive amplifier stages.

PARTS LIST

Item	Description
B1	Battery, 1.5-volt "AA", "C", or "D" cell
C1, C2, C3	Capacitor, 0.01-μF, 100-Vdc
Q1, Q2	Transistor, pnp, 2N404
R1, R3	Resistor, 100,000-ohm, ½-watt
R2, R4	Resistor, 10,000-ohm, ½-watt
S1	Switch, spst

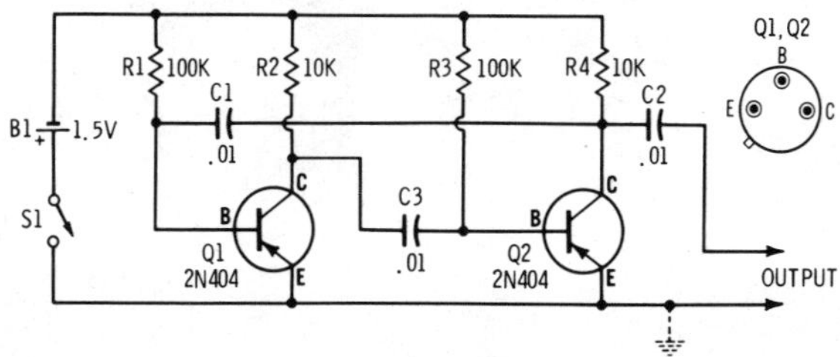

Fig. 87. Circuit for a multivibrator that produces frequencies as high as 14 MHz.

88 MODULATOR FOR SIGNAL GENERATORS

This simple four diode "bridge" allows you to easily modulate the output of rf oscillators and generators.

With the rf input to the modulator unmodulated, set the output level of the audio generator to 1/10 of the level of the rf oscillator. The signal appearing at the output terminals will be the rf signal modulated to approximately 30 percent by the af signal (30 percent is standard test instrument modulation).

PARTS LIST

Item	Description
D1, D2, D3, D4	Diode, 1N60

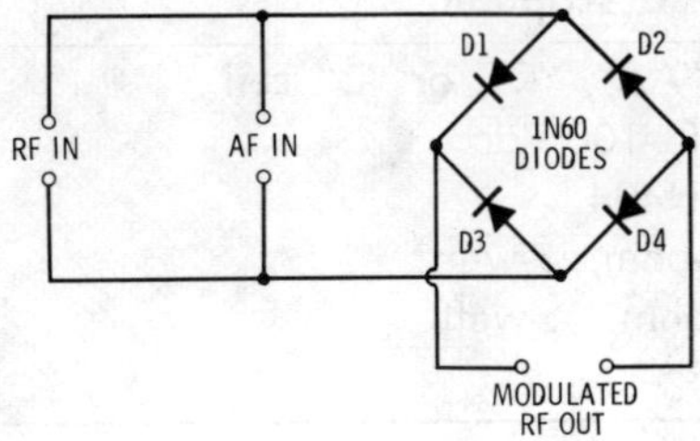

Fig. 88. Modulator for an rf signal generator.

TRANSMITTER PROJECTS

89 FM WIRELESS MICROPHONE

Just speak into the microphone, and you can broadcast to an fm receiver at distances up to 90 feet. Use standard rf wiring techniques, and construct coil L1 exactly as shown. The best speech clarity is obtained by using a crystal or ceramic microphone. For music reproduction, use a dynamic microphone.

The unit can be assembled on a perforated board using push-in terminals for tie points. The cabinet must be metal to pre-

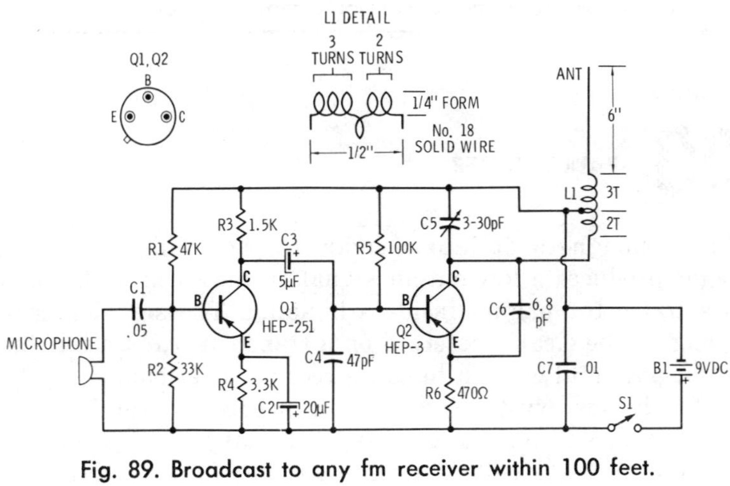

Fig. 89. Broadcast to any fm receiver within 100 feet.

vent hand capacity from changing the output frequency. Pass the 6-inch solid-wire antenna through the metal case using a ¼-inch hole and a rubber grommet for an insulator.

PARTS LIST

Item	Description
B1	Battery, 9-volt, Type 2U6 (or equiv)
C1	Capacitor, 0.05-μF, 3-Vdc
C2	Capacitor, electrolytic, 20-μF, 6-Vdc
C3	Capacitor, electrolytic, 5-μF, 10-Vdc
C4	Capacitor, 47-pF, 100-Vdc
C5	Capacitor, trimmer, 3- to 30-pF
C6	Capacitor, ceramic, 6.8-pF
C7	Capacitor,0.01-μF,10-Vdc
L1	See pictorial detail
Q1	Transistor, pnp, Motorola HEP-251 (or equiv)
Q2	Transistor, pnp, Motorola HEP-3 (or equiv)
R1	Resistor, 47,000-ohm, ½-watt
R2	Resistor, 33,000-ohm, ½-watt
R3	Resistor, 1500-ohm, ½-watt
R4	Resistor, 3300-ohm, ½-watt
R5	Resistor, 100,000-ohm, ½-watt
R6	Resistor, 470-ohm, ½-watt
S1	Switch, spst
Misc.	Microphone, crystal or ceramic element

90 RADIO PAGER

Small enough to fit into a cigarette pack, this pocket radio pager produces a low output signal on the Citizens Band that is suitable for paging inside a building. The signal is strong enough to be clearly received on a standard transceiver, but it is not powerful enough to cause receiver overload.

If only one crystal frequency is needed, socket X1 can be eliminated, and an overtone type crystal can be soldered directly into the circuit. The whip antenna is a standard three-

PARTS LIST

Item	Description
B1	Battery, 9V, Type 2U6 (or equiv)
C1, C2	Capacitor, 0.001-μF, 100-Vdc
C3	Capacitor, trimmer, 50-pF
L1	10 turns No. 16 enameled wire wound on ⅜-inch form, spaced 1-inch end to end.
Q1	Transistor, npn, Motorola HEP-50 (or equiv)
R1	Resistor, 47,000-ohm, ½-watt
R2	Resistor, 10,000-ohm, ½-watt
R3	Resistor, 330-ohm, ½-watt
X1	Socket to match crystal
Misc.	Microphone, carbon element

section walkie-talkie type. The carbon microphone can be a telephone transmitter.

To tune the pager, receive the signal on an S-meter equipped transceiver and adjust trimmer C3 for maximum output. Key the transmitter several times to make certain the oscillator "starts" each time. If the oscillator doesn't start, alter the adjustment of C3 *slightly* until dependable oscillator starting is obtained.

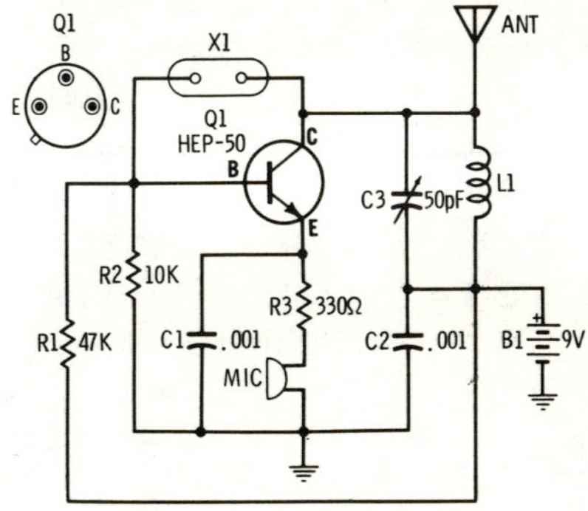

Fig. 90. This pager works on the Citizens Band.

91 MULTIPURPOSE TRANSMITTER

Utilizing 27-MHz overtone crystals, this low power transmitter can be used to provide precise frequency markers for CB dial alignment or for general receiver alignment. It can also serve as the transmitter portion of a 27-MHz radio-control circuit for camera tripping, model control, etc.

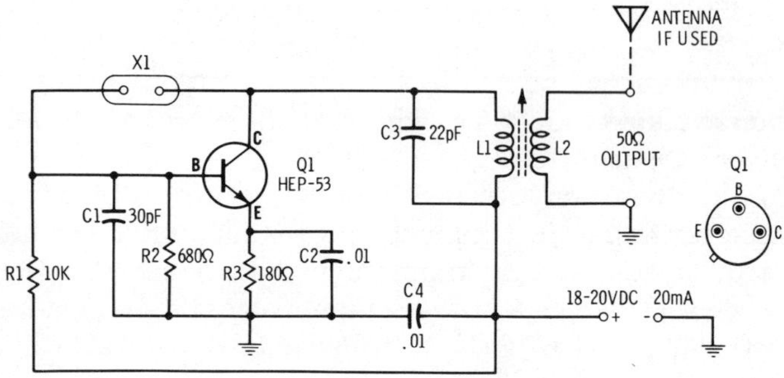

Fig. 91. Low power transmitter for operation in the 27-MHz band.

PARTS LIST

Item	Description
C1	Capacitor, 30-pF, 100-Vdc
C2, C4	Capacitor, 0.01-μF, 100-Vdc
C3	Capacitor, silver mica, 22-pF
L1	15 turns No. 22 enameled wire close-wound on a ⅜-in. powered-iron slug-tuned form (Miller 4400-3 or equiv)
L2	2 turns No. 18 enameled wire over cold end of L1
Q1	Transistor, npn, Motorola HEP-53 (or equiv)
R1	Resistor, 10,000-ohm, ½-watt
R2	Resistor, 680-ohm, ½-watt
R3	Resistor, 180-ohm, ½-watt
X1	Socket to match crystal

Coils L1 and L2 are wound on a J. W. Miller 4400-3 coil form. Attach the end of a piece of No. 22 enameled wire to the terminal closest to the mounting screw and wind 15 close-spaced turns. Push the bottom terminal against the bottom of the coil and solder.

Coil L2 is two turns of No. 18 enameled wire, wound over the bottom end of coil L1. Twist the wires of coil L2 together to secure. Finally, cover the entire coil with coil dope and allow it to dry overnight.

Plug in an overtone 27-MHz crystal and adjust the coil slug for maximum output as indicated on a receiver S-meter. The crystal frequency can be slightly changed (within a few hundred hertz) by adjustment of the coil slug.

92 15-METER FLEA-POWER TRANSMITTER

Any ham can work the world with a California kilowatt, but working out with 100 milliwatts on 15 meters is a real challenge. Use a metal chassis and good rf wiring practices when building this rig.

When cutting the Miniductor to length, cut through the plastic supports first—don't try to tear the wire through the supports.

If closing the key fails to *start* the oscillator each and every time, change the value of resistor R2 in small increments (either up or down) until you obtain reliable crystal starting.

PARTS LIST

Item	Description
B1	Battery, 9-volt, Type 912 (or equiv)
C1	Capacitor, 0.001-μF, 100-Vdc
C2	Capacitor, 0.005-μF, 100-Vdc
C3	Capacitor, variable or trimmer, 30-pF
C5	Capacitor, 0.005-μF, 400-Vdc
L1	17 turns of B & W No. 3007 Miniductor tapped at 8 turns from the battery end
Q1	Transistor, pnp, Motorola HEP-3 (or equiv)
R1	Resistor, 10,000-ohm, ½-watt
R2	Resistor, 51,000-ohm, ½-watt
R3	Resistor, 470-ohm, ½-watt
X1	Crystal, 21-MHz fundamental, and socket to match crystal
Misc.	Telegraph key

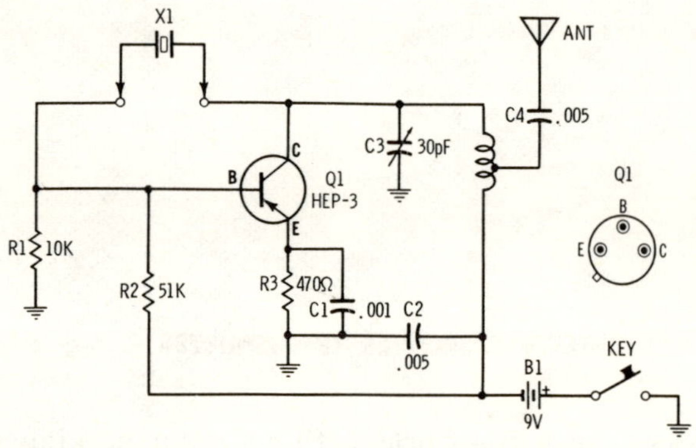

Fig. 92. This transmitter puts out 100 milliwatts on 15 meters.

MISCELLANEOUS PROJECTS

93 TREASURE FINDER

You won't find Long John Silver's buried treasure, but you might accumulate an impressive collection of bottle caps from under the beach sand. You may even find a few quarters and dimes.

This treasure finder keeps cost down by using a transistor radio as the detector. The unit is assembled on a perforated board; rigid component mounting is a *must*. It is strapped to a broom handle, close to the bottom where search coil L1 is mounted. A transistor radio is mounted near the top of the handle.

With the radio tuned to a "weak station," adjust C1 so that the treasure finder "beats" against the received station, and produces a whistle in the speaker.

When the search head passes over buried material, the inductance of coil L1 is changed by the metal, thereby changing the frequency of the treasure finder. This also changes the whistle frequency heard from the radio speaker.

The search coil consists of 18 turns of No. 22 enameled wire "scramble-wound" (which means don't be neat) on a 4-inch diameter form. This form can be a plastic tube, a wood puck, etc.—anything except metal. After the coil is wound and checked out, saturate the coil with *coil dope*. If a single loop of the coil is not firmly cemented, the unit will be unstable.

PARTS LIST

Item	Description
B1	Battery, 9-volt, Type 2U6 (or equiv)
C1	Capacitor, trimmer or variable, 280-pF
C2	Capacitor, silver mica, 100-pF
C3	Capacitor, 0.05-μF, 25-Vdc
C4	Capacitor, electrolytic, 5-μF, 10-Vdc
L1	Search coil consists of 18 turns of No. 22 enameled wire scramble-wound on a 4-in. diameter form.
Q1	Transistor, npn, GE-5 (or equiv)
R1	Resistor, 680-ohm, ½-watt
R2	Resistor, 10,000-ohm, ½-watt
R3	Resistor, 47,000-ohm, ½-watt

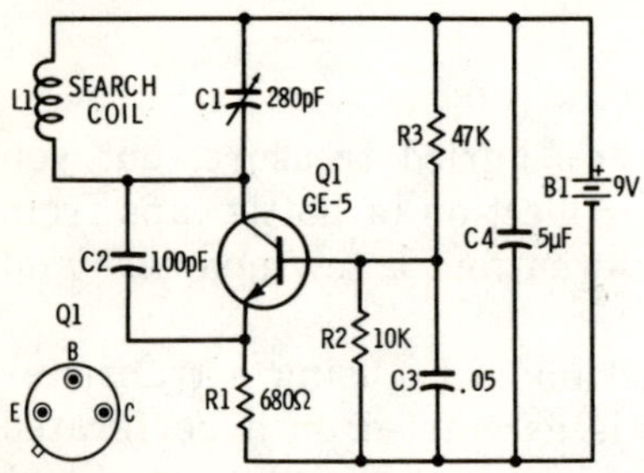

Fig. 93. This circuit can be used with a small portable radio to locate buried objects.

94 ELECTRONIC FISH LURE

"Click-click-click" might not sound like much to you, but to a fish it's the dinner bell. That's the lure of this circuit. Place the entire unit in a waterproof container such as a plastic bag, lower it over the side, and wait for the fish to hit the hook.

For proper operation, transformer T1 must be the subminiature type about half as large as your thumb. The earphone must be a crystal type such as the ones supplied with transistor radios.

PARTS LIST

Item	Description
B1, B2	Battery, 1.5-volt "AA" or "AAA" cells
C1	Capacitor, electrolytic, 50-μF, 6-Vdc
C2	Capacitor, electrolytic, 50-μF, 25-Vdc
Q1	Transistor, pnp, Motorola HEP-230 (or equiv)
R1	Potentiometer with switch, 2500-ohm, miniature
R2	Resistor, 27,000-ohm, ½-watt
S1	Switch, spst (part of R1)
T1	Transformer, subminiature transistor output, 500-ohm center-tapped primary, 3.2-ohm secondary
Misc.	Crystal earphone from transistor radio (or equiv)

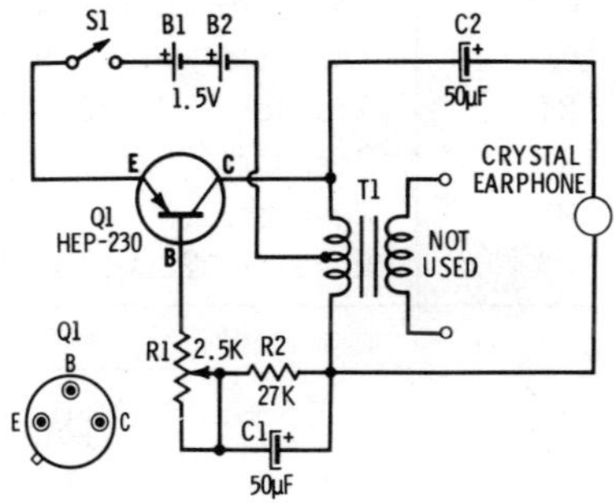

Fig. 94. Circuit for an audible fish lure.

95 OTL AMPLIFIER

OTL means *output transformerless*. Consequently, you save three dollars on the cost of this 2-watt amplifier. In addition, do not overlook wide frequency response since there is no transformer to lop off lows and highs.

The amplifier should be assembled in a metal cabinet, with the cabinet serving as the heat sink for transistor Q3. Make certain that an insulator, such as mica, is used between the case of transistor Q3—the collector connection—and the cabinet. Ensure proper heat transfer through the insulator by smearing a bit of silicone grease on both sides of the insulator.

A 4-, 8-, or 16-ohm speaker can be used although the power output decreases as the speaker impedance increases.

Because of normal transistor differences, there might be excessive distortion; if so, alter the value of resistor R4 in small increments until the distortion is reduced.

If a volume control is needed, connect a potentiometer of 10,000 ohms or higher in front of C1 as shown inside the dotted lines.

PARTS LIST

Item	Description
C1, C2	Capacitor, electrolytic, 10-μF, 6-Vdc
C3	Capacitor, electrolytic, 50-μF, 6-Vdc
Q1	Transistor, pnp, Motorola HEP-253 (or equiv)
Q2	Transistor, pnp, Motorola HEP-254 (or equiv)
Q3	Transistor, pnp, Motorola HEP-623 (or equiv)
R1	Resistor, 100,000-ohm, ½-watt
R2	Resistor, 15,000-ohm, ½-watt
R3	Resistor, 1000-ohm, ½-watt
R4	Resistor, 200,000-ohm, ½-watt

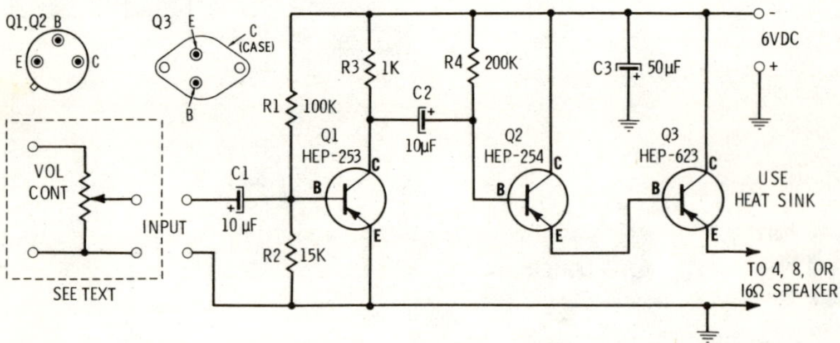

Fig. 95. No power transformer is required for this 1-watt audio amplifier.

96 BUDGET SCOPE CALIBRATOR

You can make accurate voltage measurements with your oscilloscope if you calibrate the vertical input with the scope calibrator, shown in Fig. 96A.

Just three components provide a square-wave output that varies from "0" to +5 volts. Connect the oscilloscope across zener diode D1, and adjust the vertical gain of the oscilloscope

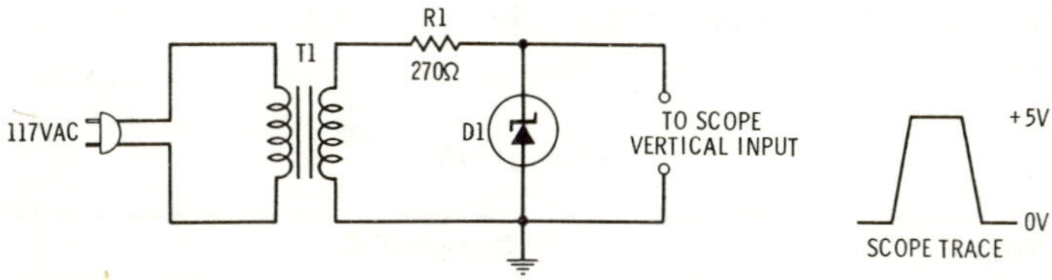

Fig. 96A. This circuit can be used to accurately calibrate your oscilloscope.

Fig. 96B. The addition of a low cost selector switch and a ten times per step resistance divider using 5% or 1% resistors will give this budget scope calibrator outputs of 5.0, 0.5, and 0.05 peak-to-peak volts. Binding posts should be spaced 3/4 inch to accept the standard instrument-type double banana plugs.

so that the square wave exactly fills one vertical division. This provides a calibration of 5 volts peak-to-peak per division. The vertical attenuator per division of the scope then provides multiples of the calibration such as 0.5 V/per division, etc. Since the calibrator varies from "0" volts, it might be necessary to adjust the vertical centering when the dc input of the scope is used.

PARTS LIST

Item	Description
D1	Zener diode, 5-volt, 250-mW
R1	Resistor, 270-ohm, ½-watt
T1	Transformer, filament, 117-to 6.3 Vac

97 TWO-SET ANTENNA COUPLER

Direct connection of two television sets to the same antenna lead can produce severe ghosting and color shifts. For the best results, the antenna connections of the two sets should be electrically isolated. You can do this with the three-resistor, two-set antenna coupler shown in Fig. 97.

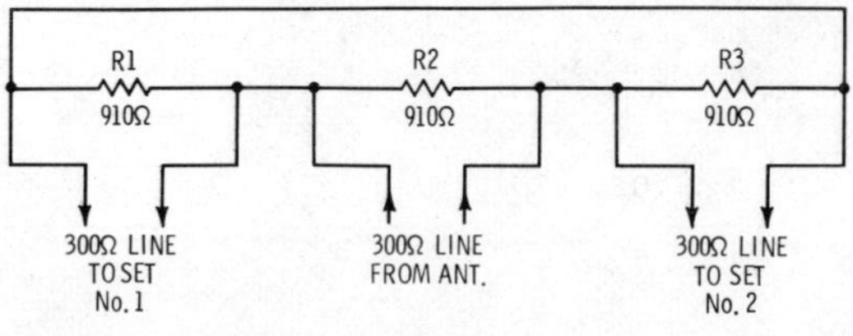

Fig. 97. Circuit for coupling two television receivers to one antenna.

PARTS LIST

Item	Description
R1, R2, R3	Resistor, 910-ohm, ½-watt
Misc.	Lengths of 300-ohm twin lead

Since there is a small signal loss in the splitting process, the signals should be moderately strong, with little or no visible "snow."

98 TWO-WAY SIGNALER

Using diode switching, a pair of wires can control two circuits that would normally require three or four wires. Although it is illustrated here with lamps, the same idea can be used for telephone circuits.

When polarity-reversing switch S1 is set so that the positive battery terminal is connected to the diodes as shown in Fig. 98, only lamp I1 lights. Lamp I1 remains off because diode D2 blocks the dc to the lamp.

PARTS LIST

Item	Description
B1	Battery, 6-volt, or four 1.5-volt "D" cells
D1, D2	Rectifier, silicon, 750-mA, 50-PIV
I1, I2	Pilot lamp, miniature, bayonet-base, 6.3-volt, 150-mA
S1	Switch, dpdt toggle

When switch S1 is thrown to the left, the battery polarity is reversed and a negative voltage is applied to the diode. Diode D2 conducts, lighting lamp I2, and diode D1 blocks the current to lamp I1.

If a carbon microphone is connected in series with the battery, and the lamps are replaced by headphones, switch S1 will determine which of the two headphones will receive the transmitted signal.

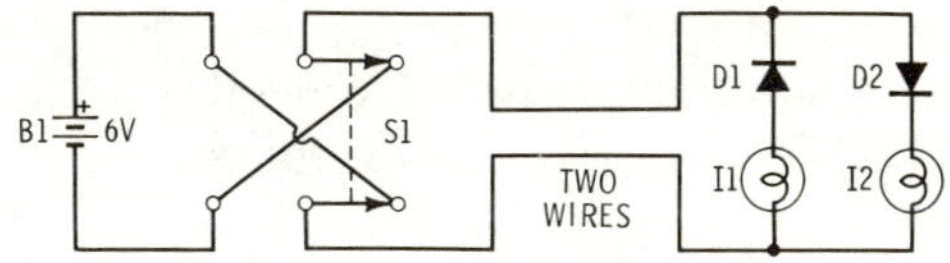

Fig. 98. With this circuit you can use two wires to transmit two signals.

99 PRECISION FREQUENCY OSCILLATOR

If you need a precision, frequency-controlled signal source for remote triggering of a telephone "snooper" or other selective device, you might try an electromechanical resonator.

The frequency of resonator F1 in the transmitter unit should match the frequency of the reed relays in the receiving unit. A switch will allow several different resonator frequencies to be installed in the transmitter.

PARTS LIST

Item	Description
C1	Capacitor, 0.2-μF, 10-Vdc
F1	Twintron resonator (H. B. Engineering Corp., 1101 Ripley St., Silver Spring, Md. 20910)
Q1	Darlington amplifier, GE D16P1 (or equiv)
R1	Resistor, 4.7-megohms, ½-watt
R2	Resistor, 2000-ohm, ½-watt
R3	Resistor, 560-ohm, ½-watt

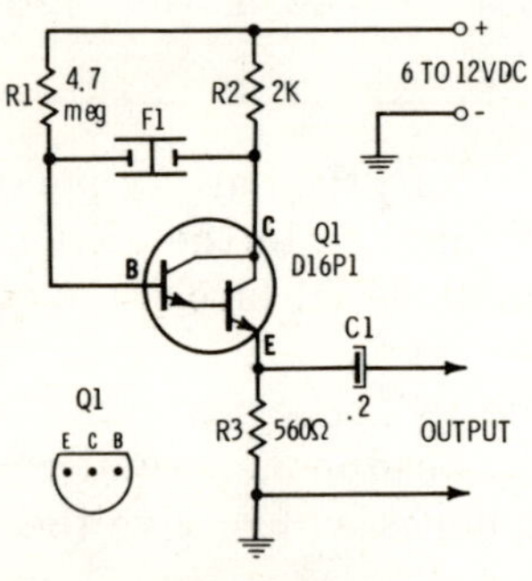

Fig. 99. This precision oscillator uses very few components.

A small amplifier with a speaker connects to the output of a darlington amplifier (Q1). In operation, the resonator passes only the tuned frequency as positive feedback, causing the amplifier to oscillate at the resonant frequency.

SCHEMATIC SYMBOLS

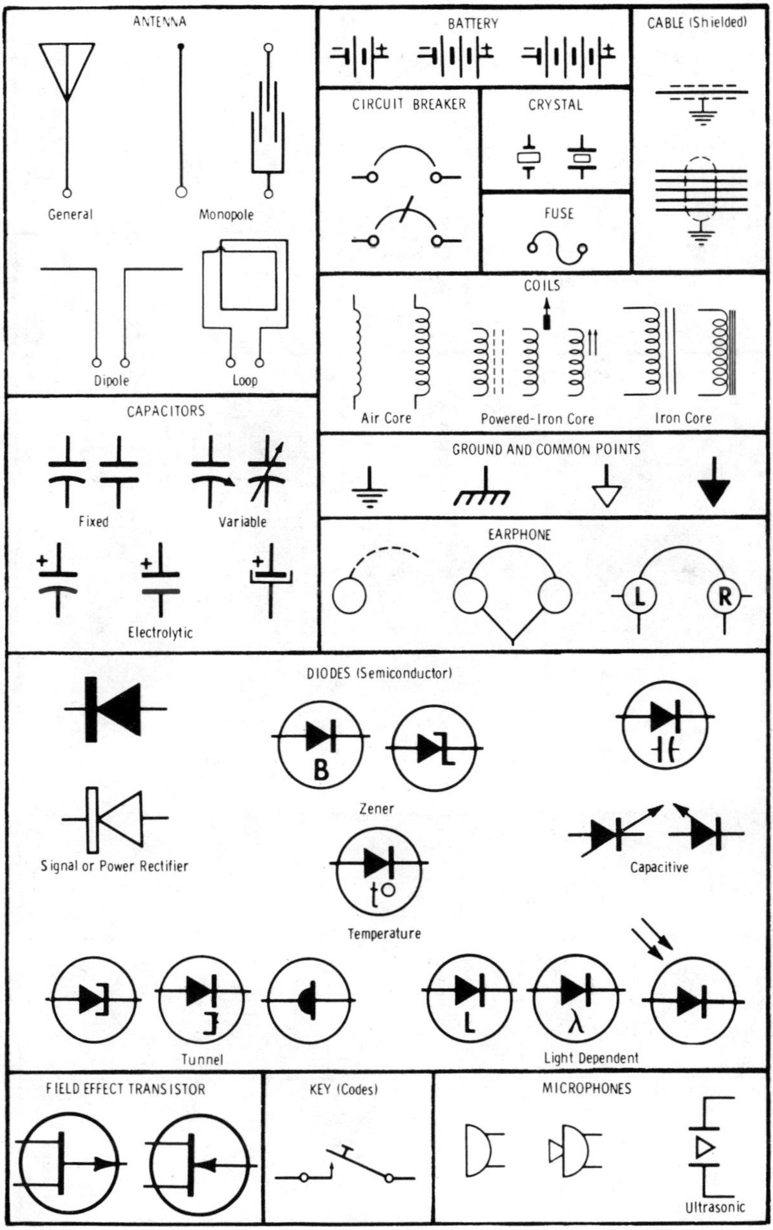

SCHEMATIC SYMBOLS

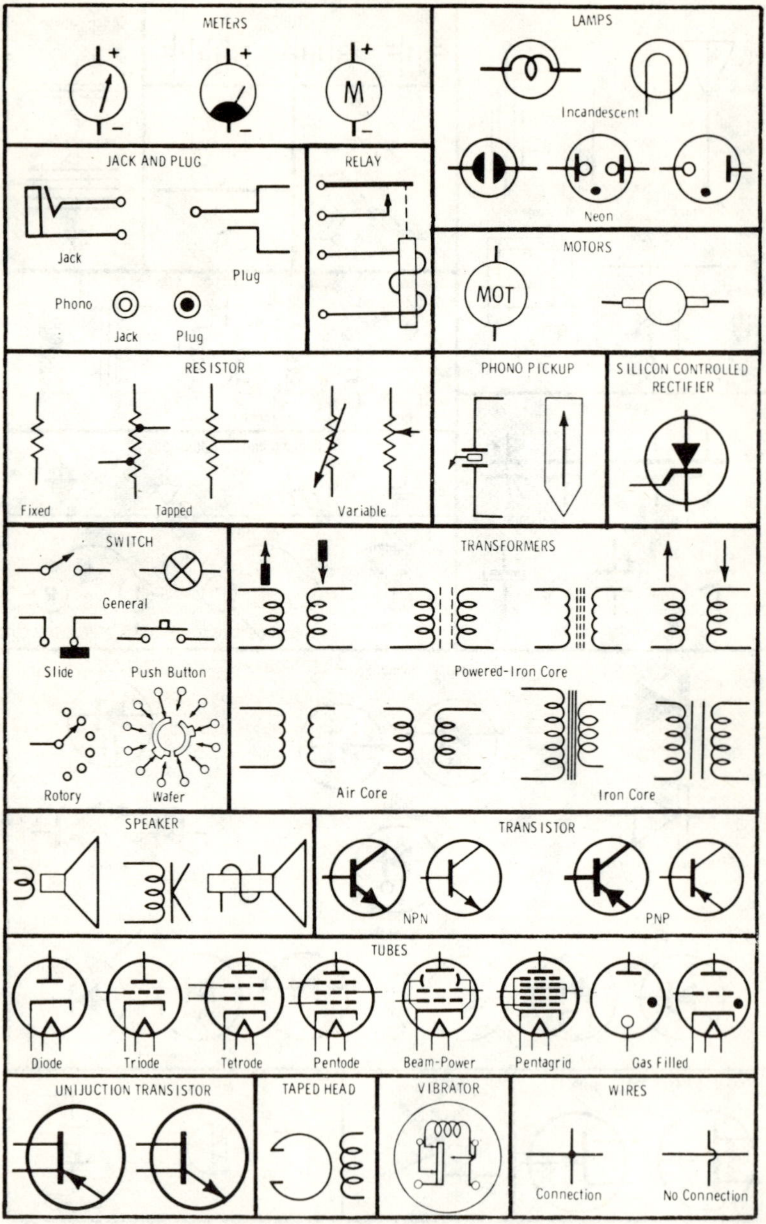